AF616363

Global Warming

Origin, Significance and Management

Global Warming

Origin, Significance and Management

R. Chattopadhyay

Mandira Chatterjee

GLOBAL VISION PUBLISHING HOUSE

20, Ansari Road, Daryaganj, New Delhi-110002 (INDIA)

GLOBAL VISION PUBLISHING HOUSE
F-4, 1st Floor, 'Hari Sadan' 20, Ansari Road,
Daryaganj, New Delhi-110002 (INDIA)
Tel.: 23261581, 23276291, 43037885 Mob: 9810644769
Email: nsingh_2004@vsnl.net, info@globalvisionpub.com
Website: www.globalvisionpub.com

Global Warming: Origin, Significance and Management

First Edition 2012

ISBN: 978-81-8220-468-3

PRINTED IN INDIA

Published by Dr. N.K. Singh for Global Vision Publishing House, New Delhi-2 and *Printed at* G.S. Offset, Naveen Shahdara, Delhi-32.

Preface

We cannot stop industries to grow, abandon mechanized transport systems or prevent growing urbanization. On the other hand, we need more of them to cater to the growing needs of the increasing population. Therefore the solution to the problem of global warming lies in managing the anthropogenic emissions to a permissible low level along with the preservation of activities in all four nature's spheres. This is 'an inconvenient truth', as Al Gore puts the facts in his widely publicized book. His was the first attempt to make the world community and the leaders' conscious about the menace of global warming. A paradigm shift in attitude followed. In order to achieve the objective of a green environment, scientific communities across the world are engaged in finding techno-ecologically viable economic solutions to tackle the problems in various spheres... The books, websites, articles covering some or other aspects of 'global warming' along with electronic & print media providing newsflashes of happenings around the world have forced the leaders and politicians to move for positive actions. However there is a need to tell the whole truth of creation and prevention in compact format, which would assist intelligent readers in this area in his or her task to save the world from the disastrous consequences of global warming. An attempt has been made to produce the compact format in this book, which is structured to address the queries of students, researchers and professionals engaged in different spheres of activities in their efforts to manage global warming.

We appreciate the support and encouragement that we have received from our son, Romik and daughter-in-law, Robin. As eye-openers, our grandsons, Ryan and Ronan deserve special thanks. Without the active support and encouragement from our younger son, Raunak, such a big project could never get completed. Our sincere thanks to Raunak. A special word of thanks to Chiraranjan for his support.

Authors would like to gratefully acknowledge the role of Dr. N.K. Singh of Global Vision Publishing House, New Delhi for the support and help in publishing the book. Thanks are also due to the staff members of the publishing house who have made printing this book possible.

Contents

PREFACE (*v*)

1. INTRODUCTION 1

SECTION I: ORIGIN AND SIGNIFICANCE

2. THE EARTH'S ATMOSPHERE 7
   - Nature and Characteristics of Atmosphere 8
   - Layers of the Atmosphere 9
   - Composition of the Earth's Atmosphere 15
   - Physical Properties of the Atmospheric Gases 17
   - Atmospheric Pressure 17
   - Atmospheric Temperature 18
   - Measuring the Atmosphere 20
   - Summary 21
3. WEATHER AND CLIMATE 23
   - Factors Determining Weather/Climate 24
   - Global Circulation of Atmosphere 27
   - Recent Extension of Tropical belt and Hadley's Circulation by Global Warming 33
   - Disasters 34
   - Summary 39
4. SOLAR BEAM AND GLOBAL HEATING 41
   - Sun's Surface 42
   - Solar Radiation 43

Solar Radiations and Radiative Transfer of Energy 44
Climate Model 46
Effective temperature of the planet by solar emission 46
Solar Constant 47
Solar Irradiance 47
Solar Luminosity 48
Solar Irradiance and Global Warming 50
Atmospheric Effects on Incoming Solar Radiation 50
Absorption of Radiation by the Atmosphere and Heating Effect 51
Reflectivity or Albedo 52
Distribution of Incoming Solar Energy 52
External Forcing and Climate Change 54
Summary 56

5. GLOBAL WARMING AND CLIMATE CHANGE 59

Global Warming and Heat Absorbing Atmospheric Gases 60
Gas Absorbing Capacity of Ocean 60
Gas Absorbing Capacity of Forests 61
Charles David Keeling and Keeling Curves 61
Keeling Curves 64
Monitoring Weather 66
Global Warming Projection 67
Natural Processes of Climate Change 67
Anthropogenic Heat 71
Anthropogenic GHGs and Radiative Forcing 72
Potential Effects of Anthropogenic Climate Change 73
Summary 73

6. GREENHOUSE GASES AND CLIMATE CHANGE 77

Discovery of Greenhouse Gases 77
Infra-red Radiation and the Greenhouse Effect 78
Greenhouse Gases, Natural and Anthropogenic 79
Properties of Greenhouse gases 80
Individual Greenhouse Gases 85

Ozone Layer Depleting Gases 91
Summary 93

7. Biome and Ecology 95

Types of Biomes 96
Summary 118

8. The Earth Systems-Spheres and Cycles 121

The Earth's Spheres- Natural and Man-made 121
Geosphere 122
Biosphere 124
Hydrosphere 125
Atmosphere 126
Anthrosphere 127
Cycles 128
Humans and the Hydrologic Cycle 137
Summary 141

Section II: Management

9. Managing Global Warming: Methodologies 145

Management Methodologies 148
Increase or Preserve Absorption Capacities of Sinks 150
Summary 154

10. Global Efforts –UN and Nations 157

Rio Earth Summit 158
Conference of the Parties (COP). 158
Climate Change Performance Index 165
UN Framework Convention on Climate Change (UNFCCC), Dec1-12, 2008, Poznan, Poland. 166
Rating Countries based on per capita Emissions 166
Framework for Mitigation Commitments 170
Opposition to Copenhagen Protocol 171
Post Copenhagen Meet of BASIC 171
2010 United Nations Climate Change Conference 172
Summary 173

11. **CARBON MANAGEMENT SYSTEM-EMISSION LIMITS** **175**
Assessment of Emission & Emission Reduction 176
Carbon Dioxide Equivalent and Carbon Units 179
Emission Units 183
CEMARS and Carbon Neutral 184
Emission limits-Kyoto Protocol 184
Kyoto Flexible Mechanisms –Carbon Projects 185
Guidelines for Emission Intensity 186
Emission Intensity as a Measure of Country's Emissivity 188
Carbon Thinking 189

12. **CARBON SEQUESTRATION AND GEO-ENGINEERING** **191**
Geosequestration 192
Carbon Capture & Storage (CCS) Technology 192
Carbon dioxide Enrichment of Flue Gas in Power Plants 193
Separation and Capturing of Carbon dioxide 195
MEA Method 196
Sequestration of Carbon dioxide 197
Comparing Offshore Geological Storage of CO_2 to Ocean Storage of CO_2 198
New Technology to Capture Carbon in Aluminum Industry 198
Carbon Sequestration by Algae 199
Carbon Dioxide Absorption by new Metal-organic Crystals 199
Summary 200

13. **FOSSILS AND ALTERNATIVE ENERGY SOURCES** **203**
Global Power Generation Scenario 204
Carbon dioxide Generation by Fossil fuel based Power Plants 205
Reduction of Carbon dioxide Emission in Fossil Fuel Plants 206
Renewable Sources for Power Generation 210
High Silt Erosion of Hydroturbines in Himalyan Regions- Nature's Revenge 213
Wind Power 215
Solar Energy 216
Biofuel 219

Advanced Gasification 223
Proposed Giant Plant in Japan 225
Seaweeds from Waste Water 225
Royal Society on Biofuel 226
Nuclear Power 227
Mini or Micro-generation Facilities 228
Summary 230

14. GHG Management in Manufacturing Industries 233

Need to Conserve Non-renewable Resources to Reduce Emission 234
Emission Management by Industries 235
Sectorial Approach 236
Summary 246

15. Transportation 249

Emission by Transport Sector 249
Emission Management in Transport Industries 250
Road Transport –Automobile 250
Zero Pollution Motors—CAV Car 254
Aircraft-aviation Industry 254
Rail Transport 257
Summary 259

16. Conservation of Biomes and Buffers 263

Global Carbon Cycle-Forests and Oceans 264
Global Carbon Emission and Flow 264
Forest Management 266
Trees/Forests and the Global Carbon Cycle 268
Forest Preservation, Kyoto and Bali Conference 268
Deforestation 269
Deforestation Processes 271
Irreversible Environmental Changes by Deforestation 273
Ocean and Global Warming 278
Summary 284

17. Agriculture and Emission Management **287**

Agriculture and GHG Emission 287
Agriculture and Land Use Change as Source of Carbon Flux 288
Summary 294

18. Waste Management **297**

Life Cycle of Product and GHG 298
Waste Management Techniques 300
Liquid Waste Management 303
Summary 305

19. Emission Trading and Business **307**

Emission Trading or Cap and Trade of Emissions 308
Kyoto Protocol and Emission Trading 308
Emission Reduction Unit (ERU) 311
Emission Trading 312
Clean Development and Joint Implementation 313
Voluntary Market 313
Voluntary Market in USA 313
Summary 321

20. Corporate Social Responsibility and Climate Change **323**

Defining Corporate Social Responsibility 324
Growth of CSR 325
CSR Global Issue 326
Guidelines for CSR 326
Potential Benefits for Corporations 331
Voluntary Vs. Mandatory CSR 332
Shift towards Green CSR 333
Truly Committed Companies on Voluntary CSR 335
CSR & Business Areas 341
Summary 341

Bibliography **343**

Index **347**

1

Introduction

The main objective of this book is to acquaint the readers on how our growing needs for energy and goods are responsible for increasing global temperature to an alarming level and how best we can manage to limit the temperature rise by using various management tools without making compromise with our basic needs.

The carbon dioxide and other greenhouse gases absorb infra-red radiation (heat) from solar beam to make the Earth's atmosphere warm.The origin of global warming *or enhanced heating of the atmosphere* has been ascribed to the steady increase in carbon dioxide in the atmosphere over the decades – a fact firmly established by Charles David Keeling (US Scientist, considered as the father of global warming).Carbon dioxide and few other gases, such as, methane, nitrous oxide, can absorb heat from sunlight and are known as 'greenhouse gases'. Greenhouse gases have high global warming potential and they can stay in the atmosphere for 100 years and more.

The most significant effect of global warming is the climate change and consequent disasters. The weather is current atmospheric conditions with respect to prevalent humidity, pressure, precipitation and temperature. Climate is an average of these conditions over the years. Atmospheric temperature plays a major role in determining the climate at a particular location.

Atmosphere circulates from equator to poles in three separate closed loops, 30 degree latitude apart, which are known as the Hadley, Ferrel and Polar cells, leading to the formation of six climate zones, such as, Tropical, Dry, Temperate, Microthermal, Polar and Highland. The climatic conditions of the zone over a long period have resulted in the formation of biome with a wide spectrum of biodiversities. With the increase in global temperature, the wind circulation cell containing hot tropical belt would expand accompanied by expansion of climate zones in this region. According to recent findings the tropical hot belt or Hadley's cell has shown an expansion by 2 to 4.5 degrees in latitude, beyond the 1979 figures. This can lead to fundamental shift in ecosystem, biodiversity and climate.

Biomes and ecosystems are created by various climatic conditions. Forests and oceans are two major biomes and also the sinks for carbon dioxide. Biomes are dynamic entities, and prolonged environmental disruptions, such as, that caused by climate change, may alter a specific ecosystem irreversibly.

Nature's spheres of activities, including carbon cycle, are confined to geosphere, biosphere, hydrosphere, and atmosphere. To these natural spheres, a more recent addition is anthrosphere or the areas of human activities. Anthrospheric emissions are mainly responsible for increase in greenhouse gases in the atmosphere.

Industries are here to stay & grow to cater to the increasing needs of growing populations and also to the changing life style in modern urban settings using power hungry devices. Therefore, the management objective in limiting emission is to strike a balance between anthropogenic emissions and industrial (GDP) growth. The successive UN- sponsored conventions on climate change are held to formulate strategies on the management of global warming. Of all the conventions held so far, Kyoto Protocol has made significant contributions in the management of global warming. Kyoto Protocol have set the goals, norms and action plans to reduce directly and indirectly GHG emissions. The signatory nations gave the commitments of time-bound programme to reduce the emission figures as stipulated in the protocol. Non signatories have resorted to voluntary emission reduction schemes. The goal is to reduce emissions from all possible sources of greenhouse gases while preserving their sinks, such as, forests, oceans and associated biodiversities. Economic incentive in terms of carbon credit has been a prime mover in reducing emissions as can be seen by the ever growing market for trading on carbon credit.

The major sources of anthrospheric emissions include power generation industries and transport systems, which are based on burning fossil fuels. Other large contributors are manufacturing industries, like, steel, cement, chemicals etc. To cater to the needs of growing population, these industries need to grow. Large-scale deforestation and pollution of ocean water have caused shrinkage in their capacities to absorb carbon dioxide. The emission management practices adopted so far include the use of cleaner energy sources, geosequestration of carbon dioxide, adoption of energy efficient processes in manufacturing, efficient agricultural practices, waste management, preservation of major sinks like forest and ocean.

Current practice of corporate undertaking voluntary social responsibilities for cutting down the emission figures in their companies would in long run help in providing the solution pertaining to the problem of global warming without cutting down the GDP.

India's race to become an economic power has propelled it to number three in the list of top carbon polluters by 2011. India's greenhouse gas emissions will keep rising as it tries to lift millions out of poverty and connect nearly half a billion people to electricity grids. Government of India would introduce soon new carbon trading scheme called, called Perform, Achieve and Trade (PAT). India is starting a mandatory scheme that sets benchmark efficiency levels for 563 big polluting from power plants to steel mills and cement plants, that account for 54 percent of the country's energy consumption.

The most important challenge of this century is how to limit global warming. In evolving a techno-ecological viable solution to this burning problem it is important to know how and why it is happening and then how to control the same. In this book we have attempted to provide answers to all the how, why and how's pertaining to this concurrent problem of global warming. None of the recent publications has such a wide coverage on this subject. It is felt that the book should be able to address to the queries of students, researchers, professionals and others, who are actively engaged in various organizations in this multi-disciplinary concurrent subject of global warming.

SECTION I: ORIGIN AND SIGNIFICANCE

- The Earth's Atmosphere
- Weather and Climate
- Solar Beam and Global Heating
- Global Warming and Climate Change
- Greenhouse Gases and Climate Change
- Biome and Ecology
- The Earth Systems- Spheres and Cycles

2

The Earth's Atmosphere

An *atmosphere* (from Greek *atmos*, 'vapor' + *sphaira*, 'sphere') or vapor-sphere is a layer of gases that surrounds the Earth and of sufficient mass for holding by the gravity. The atmosphere is one of the four spheres of the nature's activities, the other three being hydrosphere, biosphere and geosphere. Unlike the atmosphere in other planets, the presence of free oxygen, water vapor and carbon dioxide makes the Earth's atmosphere a unique entity for the existence and survival of life on the planet. The characteristic properties of the atmosphere, such as, composition, temperature and pressure determine the weather and its long term version, the climate.

The temperature, the most significant factor, forms the basis of the division of five atmospheric layers. The five layers of the atmosphere have different densities, pressures, compositions and temperatures. The ozone layer in the stratosphere filters out harmful ultra violet rays from solar beam before it reaches to troposphere, the layer close to the Earth's surface. The weather and climate are directly related to the conditions in the troposphere.

The two types of gases present in the atmosphere are termed as fixed and variable components. Fixed components consist of mainly nitrogen and oxygen and amount to 99% of the total atmospheric gases. *The fixed components (99% of atmospheric gases) have little effect on weather and climate. The variable components, constituting less than 1 percent of the atmospheric gases, have a much greater influence on both short-term weather and long-term climate.* The minor gases, such as, water vapor, carbon dioxide, methane, nitrous oxide, and sulphur dioxide absorb heat (infra-red) emitted by solar beam and thus make the atmosphere warm. These gases are known as 'greenhouse gases'. However the increase in the atmospheric greenhouse gas content beyond that produced by natural cycles can lead to an increase in global temperature to higher level than that produced by natural process. The global warming beyond certain limits can cause drastic changes in the climatic conditions across the globe, with consequent disasters.

In this chapter, the structure and properties the Earth's atmosphere and its different layers are to be discussed.

Nature and Characteristics of Atmosphere[1, 2, 3]

The present atmosphere of the Earth is not the original one, which was formed by the release of dissolved gases during cooling of molten rocks. The original atmosphere in all probabilities had carbon dioxide, nitrogen and water[1]. Without the presence of free oxygen, the atmosphere was therefore a *reducing atmosphere.* The current atmosphere contains high amount of oxygen plus inert nitrogen (neutral), and is therefore *oxidizing in nature.*

The main source of oxygen in the atmosphere is green plants. Plants use sunlight to transform carbon dioxide and water into organic matter, and release oxygen in the atmosphere. Oxygen is required by higher forms of life. The oxygen in our atmosphere was almost all produced by plants (cynobacteria) or, more colloquially, *blue-green algae*. Green plants and rain forests have converted an envelope of carbon dioxide and nitrogen into an oxygen rich atmosphere, making this planet suitable for us to live. Nitrogen is an ingredient in building proteins and nucleic acids (DNA) and essential element for living beings. Both the elements maintain same strength despite enormous amount consumed by increasing population each day due to perfect nature's cycle of recovery[2, 3].

The atmosphere protects life on earth by absorbing ultraviolet solar radiation and reducing temperature extreme between day and night. The atmospheric gases, such as carbon dioxide absorbs the infra red portion of the solar radiation. This heats up the atmosphere while allowing visible light to reflect on the Earth's surface. There is no definite boundary between the atmosphere and the *outer space*. The *Karman line*, at 100 km (62 miles or 328,000 ft), is considered as the boundary between atmosphere and outer space.

Layers of the Atmosphere[1, 4, 5]

Temperature, the most significant factor, forms the basis of the division of five atmospheric layers, viz, troposphere, stratosphere, mesosphere, thermosphere and exosphere (fig.1.1) (ref.1). The atmospheric temperature varies from one zone to another and also in each zone. The other important property is the pressure. The pressure drops uniformly across the layers from bottom to topmost layer. Atmosphere is most dense near the surface and thins out with height until it eventually merges with space. Each layer has its well-defined function, leading to the layer just above the earth's surface, with optimum conditions for survival and growth of man kind, animal, plants and all other living beings.

The characteristic functions of the atmospheric layers (fig 1.1) are discussed as follows:

i. **Troposphere (Greek word, '*tropos*' means,' to turn or change or mix):** The troposphere is the first layer above the Earth's surface. This is the layer of atmosphere under which we live. The word troposphere means *mixing*, reflecting the fact that *turbulent* mixing plays an important role in the troposphere's structure and behavior. *Weather* occurs in this layer. The *thickness of the troposphere* layer varies at different latitudes and with changing weather conditions. The thickness of troposphere beginning at the Earth's surface can vary between shallow layer of 7 km (4 miles) at the poles in summer and deeper one of up to 20 km (12 miles) at the equator, with some variation due to weather factors. The average depth of the troposphere is about 11 km (7 miles) in the middle latitudes. In the lowest depth of troposphere of few hundred meters to 2 km (1.2 miles) depending on landform, the airflow in the *primary boundary layer* is affected by the friction with the Earth's surface.

The special features of troposphere include the following:

a The temperature, humidity, pressure and precipitation of the day in the *troposphere* dictate the weather and longtime records indicate the climate. Therefore the weather and the climate of a particular geographical location is dependent on the prevalent short time and longtime conditions in the troposphere respectively.

b. Troposphere is denser than the other layers of the atmosphere above it (because of the weight compressing it), and it contains up to 80% of the mass of the atmosphere. Approximately 50% of the total mass of the atmosphere is located in the lower 5 km while the 30% in the rest of the troposphere

c. Lower bounding surface of troposphere is the Earth. Heat absorbed from reflected solar beams by the Earth's surface is mostly transferred to circulating air in troposphere. In addition to that troposphere gets the heat absorbed by carbon-dioxide and other greenhouse gases present in the air.

d. Vertical mixing in troposphere occurs due to upward movement of heated gas from the lower boundary layer. In the turbulent tropospheric layer, vertical mixing occurs due to solar heating of the Earth's surface. Solar heating of surface layer causes the warm, light air mass to move upward while the denser upper layer to move downwards. The vertical mixing in this layer limits the pressure drop at top of the troposphere to only 10% of that of the sea level[3].

e. Nearly all atmospheric water vapor or moisture in the atmosphere is present in the troposphere.

f. Combination of 'd' and 'e' results in cloud formation in troposphere and subsequent rainfall.

Vertical mixing combined with the presence of moisture in this layer results in cloud formation and thus precipitation. As the air mass gains height, the decrease in pressure leads to volume expansion and a consequent drop in temperature. The water vapor in the atmosphere tends to condense or solidify with the decrease in temperature and thus assists in the cloud formation and subsequent rainfall.

g. Temperature inversions limit or prevent the vertical mixing of air. Sometimes the normal vertical temperature gradient in the troposphere is inverted such that the air is colder near the surface of the Earth. The temperature equalization prevents the dense near surface layer to move upwards.

Such atmospheric stability can add to air pollution, such as formation of smog by preventing dispersion of pollutants, with pollutants emitted at ground level getting trapped underneath the temperature inversion.

The border between the troposphere and stratosphere, called the tropopause, is also a *temperature inversion zone.*

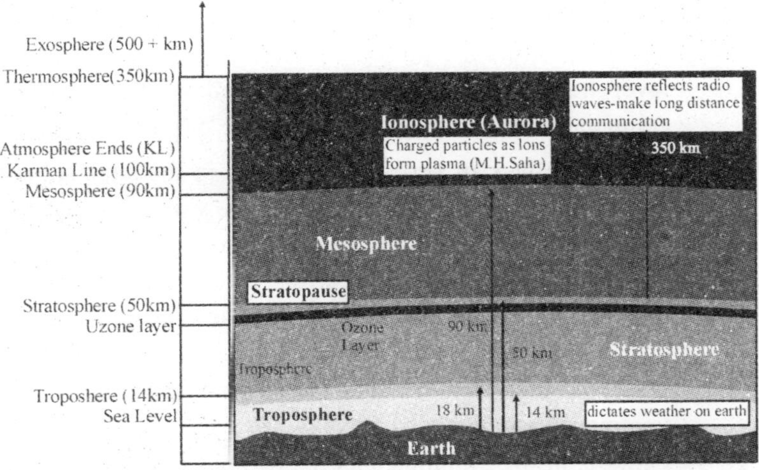

Fig1.1 is based on a diagram of layers of the atmosphere in a website (ref.4), modified by the authors. Permission obtained from the Dept. of Physics & Astronomy, Univ.of Tennessee, Knoxville, TN, USA to reproduce the figure.

ii. **Stratosphere and Ozone Layer (Latin word *'stratus'* meaning a spreading out):** The stratosphere extends beyond the troposphere, at a distance of about 50 km (160,000 ft) from the Earth's surface. The width of the stratospheric layer is in the range of about 8 and 16 km (5 and 10 miles). It varies with the latitude and the seasons. It is slightly lower in winter at mid- and at higher latitudes and slightly higher in the summer.

In *stratosphere* air flow is mostly horizontal and no turbulence. Due to stable atmospheric conditions the stratosphere is a preferred route for jet flights.

The stratosphere contains relatively high concentrations (in terms of parts per millions) of a thin 'ozone *layer*', in its lower region, approximately 15 to 35 km above earth. Ozone containing three oxygen atoms compared to two oxygen atoms in an oxygen molecule is more active than oxygen. The ozone layer is primarily responsible for absorbing the ultraviolet radiation from the Sun. By shielding the intense influx of ultraviolet rays from reaching the surface protects the life on the Earth, However the ozone layer has been found to be depleted by fluorocarbon compounds emissions of anthropogenic origin.

Due to the lack of vertical convection in the stratosphere, materials that get into the stratosphere can stay there for long times. The unreacted ozone-destroying fluorocarbons can stay beyond the ozone layers in the stratosphere for long period of 1000 years and above. Fluorocarbons belong to greenhouse gas family. The longer stay with high heat absorbing capacities of fluorocarbon can add considerably to atmospheric heat.

Large volcanic eruptions and major meteorite impacts can fling aerosol particles up into the stratosphere where they can stay for months or years. The most significant effect of aerosol is the destruction of stratospheric ozone layer.

Another source of anthropogenic greenhouse gases in the stratosphere is exhaust gas emission by rocket launches & jet planes.

The stratosphere is very dry; hence clouds are rare. However stratospheric clouds (polar stratospheric clouds or PSCs) appear in lower stratosphere at altitudes of 15 to 25 km (9.3 to 15.5 miles) near poles in winter, when temperature dips below -78° C. PSCs help formation of the infamous 'holes in the ozone layer' by supporting chlorine formation which catalyzes destruction of ozone layer. PSCs are also called *nacreous clouds.* Various types of waves and tides in the atmosphere influence the stratosphere. The waves and tides influence the flows of air in the stratosphere and can also cause regional heating of this layer of the atmosphere.

A rare type of electrical discharge, similar to lightning, occurs in the stratosphere. These "blue jets" appear above thunderstorms, and extend from the bottom of the stratosphere up to altitudes of 40 or 50 km (25 to 31 miles).

The boundary between the stratosphere and the mesosphere above is called the stratopause.

iii. **Mesosphere:** The mesosphere layer starts about 50 km (31 miles) above the ground and goes up to 85 km (53 miles) high. Most *meteors or rock fragments* burn up in the mesosphere. The lightning called *sprites* sometimes appears in mesosphere above thunderstorms. High-altitude clouds called *nocitilucent clouds or polar mesospheric cloud* can form at mesopause in the polar regions. Scientists use sounding rockets to study the mesosphere. The top of the mesosphere is the coldest part of the atmosphere. It can get down to -100° C (-148° F).

The stratosphere and mesosphere together are referred to as middle atmosphere.

iv. **Thermosphere or Ionosphere:** The thermosphere extends from mesopause to thermopause about 90 km (56 miles) to between 350 km (218 miles) above our planet. The boundary between the thermosphere and the exosphere above it is called the thermopause.

The Earth's atmosphere ends at *Kármán line*, which is an imaginary boundary line between the Earth's atmosphere and outer space, situated at an altitude of 100 Km (62.1 miles) above the sea level. Although the thermosphere is considered as a part of the Earth's atmosphere, the air density is so low in this layer that most of the thermosphere is more similar to that of outer space. The space shuttle and the International Space Station orbit the Earth within the thermosphere.

The *ionosphere* is considered as an extension of the thermosphere. Ionosphere contains less than 0.1% of the total mass of the Earth's atmosphere. In ionosphere part of the gases present gets ionized by high energy photons from solar beams, thus creating a pool of hot ionized and non-ionized atoms of gases, such as, oxygen, nitrogen and helium. Meghnath Saha[6], a physicist from India, was the first to indicate the ionization of gases leading to plasma formation in this zone. The structure of the ionosphere is strongly influenced by the charged particle or ions containing wind from the Sun (solar wind), which in turn is governed by the level of solar activity. One measure of the structure of the ionosphere is the free electron density, which is an indicator of the degree of ionization. Plasma formed in this zone has an electron density of 10^4 to 10^8/cc. The waves and tides help move energy within the thermosphere. The collisions amongst moving ionized and neutral gases can produce powerful electrical currents in some parts of the thermosphere.

Thermosphere plays an important part in atmospheric electricity and forms the inner edge of the magnetosphere An important function of ionosphere is it's capability of reflecting radio waves, thereby making long-distance radio communication possible.

The ionosphere is very thin, but it is where the aurora (the Southern and Northern Lights) occurs. Charged particles (electrons, protons, and other ions) from space collide with atoms and molecules in the thermosphere at high latitudes, exciting them into higher energy states. Those atoms and molecules shed this excess energy by emitting photons of light, which we see as colorful auroral displays.

v. **Exosphere:** Exosphere begins at 500km and goes beyond 2000km above Earth's surface in the outer space and far beyond Earth's atmosphere

Composition of the Earth's Atmosphere[1, 2, 5]

The *chemical composition of the troposphere* is essentially uniform, excepting the water vapor content. The processes of evaporation and transpiration of water vapor occur in the troposphere. The amount of water vapor decreases with fall of temperature at higher altitudes.

Table 1.1.a: Constant • Components of Atmosphere

Gases	*Percentage (volume %)*
Nitrogen	78.08
Oxygen	20.95
Argon	0.93
Neo	0.0018
Helium	0.0005
Krypton	0.0011
Xenon	0.00009

•proportions remain constant over time & locations (1,2,5)

The compositions of atmospheric gases fall in two categories, viz., a fixed and a variable composition group. The typical compositions of each group in the troposphere are given in tab.1.1a and tab1.1b[1.2.5].

The main components (tab.1.1.a) in fixed group consist of nitrogen (78.09%) and oxygen (20.95) plus less than 1% inert gas.

The oxygen and nitrogen exist in free form as a mechanical mixture but remain in the atmosphere in the fixed ratio like a chemical compound. However the *fixed components constituting the bulk of gases in the troposphere have virtually no effect on weather, climate and other atmospheric processes.*

The variable components (table 1.1b) make up far less than 1 percent of the atmosphere. However, the *variable components have a much greater influence on both short-term weather and long-term climate*. For example, variations in water vapor in the atmosphere changes relative humidity. Gases, such as, CO_2, CH_4, N_2O, and SO_2 can remain in the atmosphere for a long period of 100 years or more and absorb heat from solar beam. These gases make the atmosphere warm; creating what is known as the "*greenhouse effect.*" Without these so-called greenhouse gases, the surface of the earth would be about 30°C cooler - too cold for most of present life on the Earth to exist. Also trace amounts of gases like CO_2 sustain plant life by forming carbohydrate by photosynthesis. However *the increase in the amounts of greenhouse gases beyond that produced by natural cycles in the atmosphere can lead to increase in global temperature beyond the optimum limits, which has the potential to adversely affect the existing weather and climate pattern across the globe.*

Table 1.1.b: Variable• Components of Atmosphere

Gases	*Percentage (volume %)*
Carbon dioxide	0.038
Water Vapor	0.40
Methane	0.00017
Ozone	0.00006
Sulfur dioxide	Trace
Nitrogen oxides	Trace

•amounts vary with time and location (Ref. [1,2,5])

In addition to gases, the atmosphere also contains particulate matter such as dust, volcanic ash, rain, and snow. These are, of course, highly variable and are generally less persistent than gas concentrations, but they can sometimes remain in the atmosphere for relatively long periods of time. The stratosphere is very dry; and the air contains little water vapor. Above stratosphere is the thermosphere, in which the air density is so low that it can be considered as a part of as outer space. Ionosphere, a part of upper thermosphere, contains atoms and ions, formed by intense solar radiation instead of gas molecules present in spheres at lower altitudes.

Physical Properties of the Atmospheric Gases[3]

The relation amongst the four basic physical properties, such as, pressure, volume, mass and temperature of the gases can be expressed by combining Boyle's and Charlie's laws as follows:

PV = RmT, where P = pressure, V = volume, m = mass, R= Gas constant, & for dry air =287 $J.kg^{-1}.K^{-1}$, and T = degree Kelvin. By substituting, density as ñ = m / V, the above equation becomes **P = RñT.** At any given pressure, increase in temperature leads to a decrease in density.

Atmospheric Pressure

Atmospheric pressure is the force per unit area that is applied perpendicularly to a surface by the surrounding gas, and is determined by a planet's gravitational force in combination with the total mass of a column of air above a location. Atmospheric pressure is generally measured in millibars (mb). Other units commonly used, include, bars, atmospheres, or millimeters of mercury. Unlike temperature, pressure decreases exponentially with altitude, as shown in fig.1.2. The highest pressure at the bottom of troposphere or sea level ranges from about 960 to 1,050 mb, with an average of 1,013 mb gradually decreases with increasing altitude to a low value beyond around 18 km height. Traces of the atmosphere can be detected as far as 500 km above the surface of the earth. The atmospheric gas pressure is related to density. At the top of Mt. Everest, pressure is as low as 300 mb, which is approximately 1/3rd of that at sea level. Therefore the available gas, (including oxygen) at this height would be one-third of that present at the sea level. The mountain climbers get more severe shortness of breath with the height, due to less oxygen.

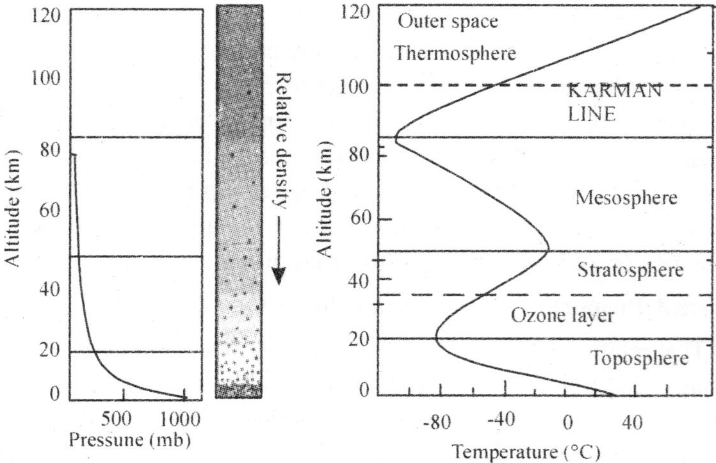

Fig. 1.2 Variation in Pressure and Density in Different Atmospheric Layers

Fig. 1.3. Variation in Temperature in Different Atmospheric Layers

Air is roughly a thousand times thinner at the top of the stratosphere than it is at sea level. Because of this, jet aircraft and weather balloons reach their maximum operational altitudes within the stratosphere.

Atmospheric Temperature

The atmospheric temperature varies from one zone to another and also in each zone as shown in fig1.3[2]

Troposphere

The average atmospheric temperature on the earth's surface is around 28 °C. Within the troposphere, temperature decreases with altitude at a rate of about 6.5° C per kilometer, and at the end of troposphere (18Km), the average temperature becomes (-) 80 °C. Mt. Everest at 856m high extends around halfway through the troposphere. Assuming a sea level temperature of 26° C (80° F), the calculated temperature on the top of Everest would be -31° C (-24° F). In fact, the temperature at Everest's summit averages -36° C, whereas temperatures in New Delhi (India) at elevation of 233 m, average to 28° C.

Sometimes the temperature does not decrease with height in the troposphere, but increases. The normal vertical temperature gradient is inverted such that the air is colder near the surface of the Earth. Such a situation is known as a *temperature inversion*. Temperature inversions limit or prevent the vertical mixing of air. This can occur when, a warmer, less dense air mass moves over a cooler, denser air mass. An inversion also occurs if radiation from the surface of the earth exceeds the solar radiation, which is common at night, or during the winter when the angle of the sun is very low in the sky.

Stratosphere

In early 1900s that radiosondes revealed a layer, about 18 km above the surface, where temperature abruptly changed and began to increase with altitude. The discovery of this reversal led to division of the atmosphere into layers based on their thermal properties. The cause of the temperature reversal is the presence of ozone layer at the bottom of stratosphere. Ozone's ability to absorb incoming ultraviolet (UV) radiation from the sun results in the temperature rises upward through the stratosphere, from (-) 80 °C to (+) 30 °C at the end of stratosphere, *i.e.*, 50Km height. This is exactly the opposite of the behavior in the troposphere where temperature drops with increasing altitude. Because of this temperature stratification, there is little convection and mixing in the stratosphere, so the layers of air there are quite stable.

Mesosphere

Above the stratosphere, temperature begins to drop again in the next layer of the atmosphere called the mesosphere (Fig1.2). This temperature decrease results from the rapidly decreasing density of the air at this altitude. The temperature of 30 °C at 50 km attitude, drops down to the coldest temperature in the Earth's atmosphere of about -90° C (-130° F) near the top of this layer.

Thermosphere

Finally, at the outer reaches of the earth's atmosphere, the intense, unfiltered radiation from the sun causes molecules of oxygen

(O_2) and nitrogen (N_2) to break apart into ions. The release of energy from the ionization causes the temperature to rise again in the thermosphere. The thermosphere extends to about 500 km above the surface of the earth. In this zone, at altitudes from 90 km to 100 km the temperature increases from (-) 100 °C to (-) 50 °C, and then rapidly increases with altitude to 500 °C and above (up to 1500 °C), depending on the solar activities.

Exosphere

The temperature remains at 1500 °C and above.

Measuring the Atmosphere

The measurements of temperature and pressure not only help in predicting the weather, but also help in forecasting long-term changes in global climate.The two most important instruments for taking measurements in the earth's atmosphere, developed hundreds of years ago, include thermometer (Galileo,1593) to measure temperature, and barometer (Evangelista Torricelli,1643) for measuring pressure. The pressure and temperature measurements at the earth's surface, were latter replaced by kite-mounted instruments and then unmanned balloons for measurements at higher altitudes. This was followed by use of the radiosonde with a radio transmitter among its many instruments, allowing data to be transmitted from the balloons no longer needed to be retrieved. A radiosonde network was developed in the United States in 1937, and continues to be in use this day under the auspices of the National Weather Service.

Technological developments has led to the current practices of continuous satellite monitoring of the atmosphere and Doppler radar enable forecasting of the impending rain or storm. A *weather radar* is a type of radar used to locate precipitation, calculate its motion, estimate its type (rain, snow, hail etc.), and forecast its future position and intensity. Modern weather radars are mostly pulse-Doppler radars, capable of detecting the motion of rain droplets in addition to intensity of the precipitation. Data can be analyzed to determine the structure of storms and their potential to cause severe weather situations.

Summary

i. Five layers of atmosphere include troposphere, stratosphere, mesosphere, thermosphere and exosphere, and the last one merging into outer space.

ii. Troposphere is the first layer above the surface and contains half of the Earth's atmosphere. It contains 80% of the total mass of the atmosphere. The presence of nitrogen & oxygen, together amounting to almost 99% has no role to play in making the weather. The major source of heat absorption is the trace amount of carbon dioxide present in the atmosphere. Nearly all atmospheric water vapor or moisture in the atmosphere is present in this layer. Weather occurs in this layer. Surface heating by solar beam leads to vertical movement of warmer lighter near surface air upward in this layer.

iii. Next layer, stratosphere contains ozone layer, which filters out the harmful UV-rays from solar radiation.

iv. *No definite boundary between the atmosphere and the outer space exists.* It slowly becomes thinner and fades into space. The 'Karman line', at 100 km (62 miles or 328,000 ft), is considered as the boundary between atmosphere and outer space.

v. The pressure of an atmosphere decreases with altitude due to the diminishing mass of gas above each location.

vi. With altitude the temperature decreases in troposphere. Ozone layer in stratosphere makes the temperature reversal, leading to increase in temperature with altitude. The discovery of this reversal led to division of the atmosphere into layers based on their thermal properties. The temperature decreases with the altitude in the next layer, *i.e.*, mesosphere, followed by monotonous increase in thermosphere.

REFERENCES

1. Michael Tennsen, *Global Warming*, Alpha Books, NY, USA, 2004
2. David D. Kemp, *Exploring Environmental Issues –An integrated approach*, Routledge, London & New York, 2004
3. Roger G. Berry and Richard J. Chorley, *Atmosphere, Weather & Climate*, Routledge, London and New York, 7th edition, reprinted in 2002, pp. 1 -18.
4. University of Tennessee, Dep.of Physics and Astronomy, The Earth's Atmosphere, http://csep10.phys.utk.edu /astr161/ lect/earth / atmosphere.htm
5. Anne E. Egger, *Earth's Atmosphere-Composition and Structure, Visionlearning*, The National Science Foundation, Arlington, Virginia, USA, in the website www.visionlearning .com/ library/ module_viewer.php?mid=107
6. Meghnath Saha, *On Ionization in the Solar Chromosphere*, Philosophical Magazine 1920. http://www.encyclopedia.com/topic/ Meghnad_Saha.aspx

3

Weather and Climate

Weather occurs in troposphere, the near surface layer of the Earth's atmosphere. *Weather* is the current atmospheric conditions with respect to prevalent humidity, wind, precipitation, and temperature in the troposphere at a particular place. *Climate* is an average of the weather over a long period. The humidity, temperature, precipitation and pressure of the atmosphere are the key factors, which determine the climate. Water vapor in humid atmosphere forms clouds, and the rainfall depends on type of cloud. The atmospheric temperature affects humidity, pressure and precipitation and thus plays a major role in determining the climate in a particular location. The temperature of a given atmosphere is determined by the amount of solar beam it receives from the sun. Solar beam striking at 90° at equatorial region produces maximum atmospheric temperature. With increasing latitude the distance between sun and earth's surface increases resulting in six different temperature or climate zones from equator to poles.

From equator to poles there are three distinct patterns of wind circulation cells, which are around 30° apart in latitude from each other in both sides of the equator. The wind circulation loops are driven mainly by the temperature gradients across the loop. The wind circulation within a loop tends to equalize the average temperature. The three wind circulation cells receive different quantum of heat from the solar beam. The increase in temperature in a loop leads to extension of the circulatory cell, thus affecting the other two cells. Thus the atmospheric temperature rise due to excess greenhouse gas emissions in one region and consequent rise in temperature can be felt across the globe.

The atmospheric temperature rise depends, amongst other factors, on the capabilities of the greenhouse gases to absorb heat (infra-red) from solar beam. Recent observation of the expansion of the tropical zone (Hadley's circulation cell) by few degrees has been ascribed as due to the increase in the amount of heat absorbing gases in the atmosphere.

As the Earth rotates on its tilted axis (23.5°) around the sun, different parts of the Earth receive higher and lower levels of radiant energy at different periods. This creates the seasons. Apart from normal seasonal variations in weather, the change in wind circulation resulting from major fluctuations in temperature and pressure, can cause abrupt changes, such as, storms, cyclone, tornadoes, hurricanes etc. The temperature changes can affect the ocean circulatory system leading to disasters like,El Nino and La Nina. Global temperature rise can result in more frequent occurrences of these disasters.

Factors Determining Weather/Climate

Humidity

With increasing temperature, solid ice (form at 0°C) transforms to liquid water followed by vapors (form at 100° C) at higher temperature. The water vapor (gas phase) in atmosphere can move freely just like any other gas. It can form clouds, fog or cause higher humidity at higher concentrations. The amount of water vapor retained by atmosphere, compared to its retention capacity, is expressed as *relative humidity* in weather reports. The relative humidity increases with more water vapor in the atmosphere or with a decrease in temperature, when vapor retention capacity decreases.

Cloud

The water vapor in the humid atmosphere needs to get together, or coalesce, in order to form clouds as a first step to precipitation. As such water vapor molecules do not attract each other to form clouds or rain droplets. Particles in the atmosphere act as *condensation nuclei* around which cloud droplets can form. These particles typically originate from ocean spray, forest fires, volcanoes and smoke stacks. This is one reason why it rains a lot more near polluting factories. The four types of clouds are as follows:

i. *Cirrus:* occur at high altitude, where air is so cold that clouds are made from thin ice crystals, resulting in the *wispy effect.*

ii. *Stratus:* flat and form closer to ground. They result in *drizzles.*

iii. *Cumulus*: These are puffy, un-threatening, or gathering, rising and *thunderous*.

iv. *Nimbus*: These are responsible for making *rain.*

A combination, like, *cumulonimbus,* consisting of cumulus plus nimbus clouds, contains rain and is responsible for making rain.

The elevation of a cloud indicates the amount of precipitation, and the possibility of storms or thunderstorms. For example, high clouds (>20,000ft) indicate arrival of storms, while clouds that are between 6,500 to 20,000ft indicate approaching rain, and low level clouds (< 6500ft) cause rain or snow. Clouds that move vertically upwards cause thunderstorms, but do not result in heavy rains.

The cloud formation and rainfall, aided by ocean and forest, are part of nature's hydrological cycle, an essential system for sustaining life on earth. In addition to making rain, clouds reflect solar beams, thereby cooling the atmosphere that lies below it. However, clouds also trap the light reflected from earth, thus heating up the atmosphere.

Precipitation

When relative humidity reaches 100%, water vapor from the atmosphere can precipitate and fall to the earth as rain, snow, hail, and sleet. Water vapor in liquid or solid form is known as *precipitate* and the formation process as *precipitation.* The form of precipitation depends on the degree of cooling.

i. ***Rain:*** For rain to occur, in addition to 100% humidity, the temperature must fall causing reduction in retention capacity. This requires movement of the vapor forming clouds upward in the troposphere, where the temperature becomes lower with higher altitude. Rain depends also on the velocity at which the atmosphere moves upwards. With slow upward movement, moisture laden atmosphere can cause light drizzle, while rapid movement can cause torrential downpour, and still faster build-up of clouds can cause thunderstorms and tornadoes. The emitted gases like sulfur dioxide or nitrous oxides from factories and cars react with water vapor in the atmosphere causing acid rain, which destroys forests, poisons water, and wears away stones.

ii ***Snow:*** The snow is the frozen precipitation resulting from the growth of ice crystals from water vapor in the Earth's atmosphere. At temperatures above -40°F (-40°C), moisture begin to form snow around dust particles or other nucleation sites and subsequently precipitates as snow. Snow crystals often grow in the supersaturated environment provided by a cloud of super cooled droplets.This process of precipitate formation is known as *Bergeron-Findeisen process*.

Pressure

The rate of upward velocity of clouds depends on pressure. The normal atmospheric pressure as measured by barometer is 760mm (30 inch) at room temperature. Higher elevations have lower pressures thus causing air to move upwards. However, downward movement of air or 'falling' air produces desert. For example, in Washington and Oregon, rising air and cloud over the Cascades results in rainfall creating dense forest area along the coast, while the falling dry air on the opposite side makes Desert Island.

Temperature

The increase in atmospheric temperature is due to the heat producing long wave length infra-red spectrum of solar radiations, which are absorbed by atmospheric gases. The increase in temperature can cause more vapor phase formation and vapor retention capacity. Lower temperature reduces vapor retention capacity thus increasing relative humidity. Also atmospheric pressure decreases with the increase in temperature and increases with the decrease in pressure. The major variables affecting weather, humidity and pressure are dependent on temperature. Temperature has therefore a profound effect on climate change.

Global Circulation of Atmosphere

This large-scale movement of air (together with ocean circulation) is responsible for decreasing thermal gradients across the Earth's surface, from the equator to the poles. In the absence of atmospheric circulation, average winter temperatures at the poles would be around -100°C rather than -30 °C as at present. Also without the atmospheric circulation, warming of one region by excessive greenhouse gas emission would have no effect on the other regions. On the otherhand, the effect of excessive rise in regional temperature would have disastrous consequences on local weather.The global warming is the average atmospheric temperature rise across the globe. The temperature averaging effect of air circulation makes the atmospheric warming as truly global in nature. Anthrospheric emissions by industrialized countries causing global warming affect nations across the globe, including the non-emitters.

Planetary rotation results in the development of three air circulation cells in each hemisphere rather than a monolithic one. The simple model incorporates three distinct patterns of wind circulation cells, from equator to poles, & are known as the *Hadley, Ferrel and Polar cells* (fig1.1). In the 3-cell model, due to the Earth's rotation, another factor, called Coriolis effect, causes moving wind to swing to the right in the northern hemisphere and to the left in the southern hemisphere, thus becoming westerly and easterly, rather than blowing north and south. Modern version of the atmospheric circulation takes into account the non-uniform nature of the Earth's surface, such as, the land and water with differences in thermal properties causing significant difference in temperature and thus producing a series of pressure cells rather a single belt in the simple model[3].

The driving force for circulation is the temperature gradient. The vertical lifts in the air circulation systems are confined to troposphere, the Earth's atmosphere responsible for climate. The vertical lift can be as high as 15 km in equatorial region and gradually decreases to around 8 km at the polar region, keeping in conformity with the change in altitudes of troposphere.

1. Hadley Cell

Hadley cell is the air circulation loop between equator and approximately 30° latitude apart in both Northern and Southern hemispheres.This area of greater heat acts as zone of *thermal lows* and known as the *intertropical convergence zone (ITCZ)*). Thermal low is an area of low pressure due to the high temperatures caused by intensive local heating in deserts and other land masses in this zone. Large-scale thermal lows help drive monsoon circulations. Equator remains the warmest location on the Earth.The wind coming from the east is called *'trade winds'*. Wind actually travels from the north east in northern hemisphere and from the south east in southern hemisphere.

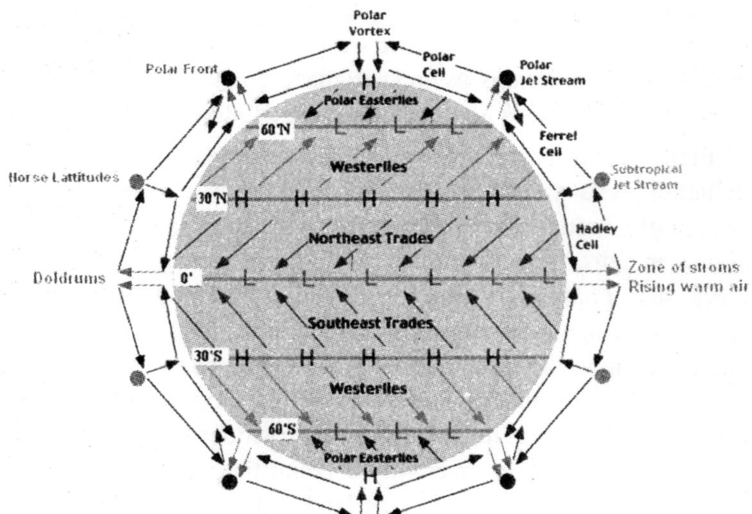

Fig. 2.1 Global 3-Cell Surface and Upper Atmosphere Circulation Pattern

The convergence of circulating trade winds from both hemispheres at the equator leads to low pressure in this region. Low pressure and rising warm air cause storms and rains around the equator. When the subtropical air reaches the *equator* the air rises vertically upwards to the top of troposphere (15km) by means of convergence and convection. It then begins to flow horizontally to the North and South poles, and thus tends to level the heat difference between the equator and poles in the circulatory cell region.

At about 30° latitude, near the altitude of tropopause, *Coriolis force* cause deflection of the polewards moving air in the Hadley cell to the right in the Northern hemisphere and to the left in the Southern Hemisphere, creating the *subtropical jet streams* that flow from west to east. The zonal flow also causes the accumulation of air in the upper atmosphere. Some of the air in the upper atmosphere sinks back to the surface (30° latitude), creating a *subtropical high pressure zone.* This zone has no rain, and plenty of sunshine, thus creating world's great deserts at 30° lattitude. From this zone, the surface air travels in two directions. A portion of the air moves back toward the equator completing the circulation system in the Hadley cell.

2. Ferrel cell

Ferrel wind circulation loop lies between approximately 30° and 60° latitudes in both the Northern and Southern Hemisphere. It is believed the cell is a forced phenomenon, induced by interaction between the Hadley and Polar cell. The stronger downward vertical motion and surface convergence at 30°N coupled with surface convergence and net upward vertical motion at 60°N induces the circulation of the Ferrel cell. This net circulation pattern is greatly upset by the exchange of polar air moving southward and tropical air moving northward.

This moving wind gets deflected by the *Coriolis Effect,* between 30° and 60° latitude towards west are called are called *prevailing westerlies.* Westerlies collide with cold air traveling from the poles.

This collision results in *frontal uplift* and the creation of the *subpolar lows* or *mid-latitude cyclones.* A small portion of this lifted air is sent back into the *Ferrel cell,* after it reaches the top of the troposphere. Most of this lifted air is directed to the *polar vortex* where it moves downward to create the *polar high.*

3. Polar cell

In the poles, winds come from the east and are called the '*polar easterlies*'. The *Coriolis force* causes them to arrive at slight angles. Surface air flow is from the poles to the equator. When the air reaches the equator, it is lifted vertically by the processes of convection and convergence. When it reaches the top of the troposphere, it begins to flow once again horizontally. However, the direction of flow is now from the equator to the poles. At the poles, the air in the upper atmosphere then descends to the Earth's surface to complete the cycle of flow.

Seasons

The Earth rotates about its axis, with a tilt at 23.5 degrees. As the Earth rotates on its tilted axis around the sun, different parts of the Earth receive higher and lower levels of radiant energy. This creates the seasons. Four seasons in western world are known as spring, summer, winter and fall. Tropical regions may have one or two more seasons, including the rainy season. The six normally occurring seasons in India (excepting South with two rainy seasons) are known as grishma (summer), varsha (rainy season), sharat (fall), hemanta (prewinter), sheeth (winter), vasamta(spring).

Climate zones[1, 4]

Areas with consistent climates are grouped together in *climate regions.* Climate influences *ecosystems,* which are the communities of plants and animals. Specific associations of organisms therefore often characterize many climate regions. The original classification by Waldimir Koppen, which is known as the *Köppen Climate classification* System, divides the Earth's surfaces into five major climate types, based on the annual and monthly *average of temperature and precipitation.* The five major climate types, are known as, Tropical (A), Dry (B), Temperate(C), Cold (D) and Polar (E).

Koppen's system is the most widely used and forms the basis of other classification systems used today. Köppen's division of the Earth's surface into climatic regions based on temperature and humidity has generally *coincided with world patterns of vegetation and soils.*

The *Koppen-Geiger classification* system, a modification of Koppen classification, recognizes six climate regions on the basis of average monthly temperatures, average monthly precipitation, and total annual precipitation values. The six climate regions are as follows:

i Tropical /Megathermal climate(**A**)

ii Dry/Arid (or Semiarid)climate (**B**)

iii. Mesothermal(**C**)/[Koppen's Temperate]

iv. Microthermal/Continental climate [Koppen's Cold] (**D**)

v. Polar climate (**E**); and a new one

vi. Highland /Alpine climate (**H**); which is not in Koppen's list.

Two subgroups, **S** - semiarid or steppe, and **W** - arid or desert, are used with the B climates. Further subgroups are designated by a second, lower case letters, such as, ***f, m, s, & w*** which distinguishes specific seasonal characteristics of temperature and precipitation. To further denote variations in climate, a third letter was added to the code, such **a, b, c, d, h &k.** The different climate zones alongside with the climatic conditions, biomes and global positioning are shown in tab.2.1.

Table 2.1: Classification of Climate Zones, Parameters, & Global Positioning

Type	*Subtype*	*Climate Parameters*	*Global positioning*
A(tropical) Low-altitude Climate	Af(tropical moist) Rainforest	RF>250mm(1" T=27°C H=77-88%	Amazon basin, Congo basin, Sumatra to New Guinea
A	Aw (West-dry Tropical)	RF=25mm(0.1") T=16°C	India, Indonesia, West & South Africa, S. America, Australia (north coast)
B (Dry)	BW (dry Topical) Desert Of earth's Surface	T=16°C P=0.25cm Covers 12%	USA(SW), Mexico(north), Argentina, Africa (north, south), Australia (central)
B Mid-Altitude Climate	BS-Steppe	T=24°C P=10cm	USA(western north), Eurasian interior, Eastern Europe to Gobi desert
C (Temperate/ Mesothermal)	Cs, Mediterian, Chaparal biome	T=7°C, P=42cm	Africal (central & south), mediterrian coast, Australia (west & south coast), Chlian coast, Cape Town
D (Coold, Mesothermal) High altitude climate	Dfc, Tiaga biome	T=(-)25 to 16°C P=31cm	Alaska, Canada, Yukon regions to Lbrador, Eurasia-from north Europe to Siberia to Pacific ocean
E (Polar)	Tundra climate, tundra biome	T=(-)22 to 6 °C.P=20cm	North America (aectic zone & Hudson Bay region), Greenland coast, South Sibera
H (Highland	Alpine biome	T=(-) 18 to 10°C P= 25cm	Rocky mountain (N. America), Andean (S. America), Alpine (Europe), Kilmanjaro (Africa), Himalya (India). Mt. Fuj I (Japan)

Note T=Temperature, RF = Average annual rainfall, P= Average annual precipitation

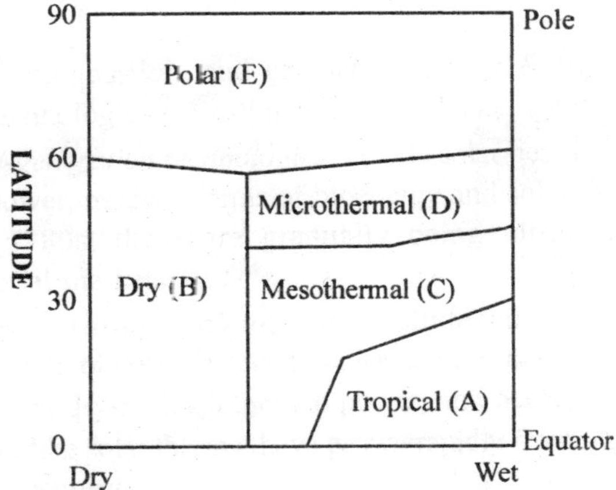

Fig. 2.2. Relationship between climate regions and geographical locations

The relationship between climate zones and their geographical locations are shown in fig.2.2. The climate zones and biomes in different regions have distinct temperature, humidity and precipitation and are maintained by the patterns of wind movements. Below polar region, the climate zones are basically divided into two, dry and wet areas. The wet area has three zones based on temperature, viz., tropical, mesothermal and microthermal with increasing latitude from equator to poles (fig.2.1).However the rise in temperature due to global warming has a profound effect in making changes in the climate zones.

Recent Extension of Tropical belt and Hadley's Circulation by Global Warming

A recent finding[5] indicates that the tropical belt covering the Hadley's circulation has extended by 2 to 4.5° in latitude (NS), *i.e.*,145 to 330 miles (225 to 530km), beyond the 1979 figures of 23.5°(NS) in latitude (Fig.2.3).Global warming and depletion of ozone layer are found to be responsible for extensions of Hadley's circulation cell. The expansion in Hadley circulation cell has resulted in the extension of tropical belt by 2 to 4.5° in latitude (NS) in Ferrell cell. The net effect is the rise in average global temperature. The major climate types in the tropical belt from 23.5° N to 23.5°S latitudes are as follows (fig.2.3).

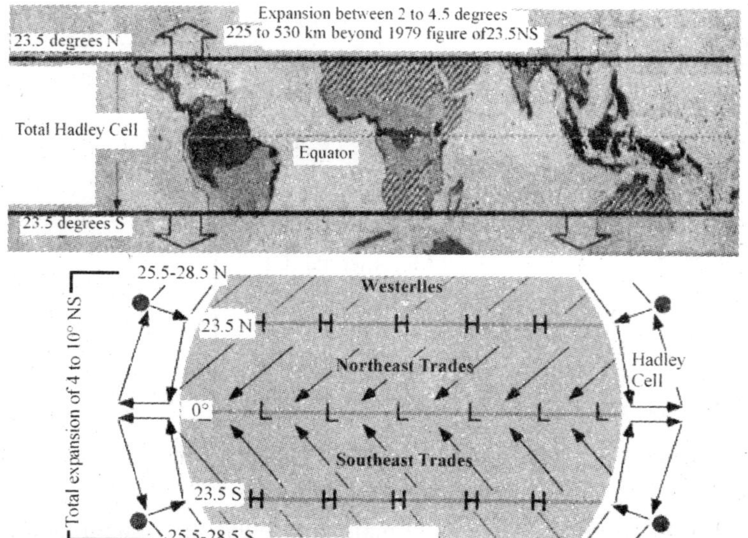

Fig. 2.3 Expansion of Hadley's circulatory cell in Tropical belt due Global Warming

The shift shall result in climate changes not only the tropical belt but also in mesothermal and microthermal zones. India's plane lands belong to tropical belt and the extension in northern latitude would cover the highland zone, such as Himalayan regions in tropical belt. The possible effects on the climate shall include, an expansion of drier and hot climate areas and a fundamental shifts in ecosystem and human settlements.

Disasters

Apart from normal seasonal variations in weather, the change in wind circulation resulting from major fluctuations in temperature and pressure, can cause abrupt changes leading to disasters, such as, storms, cyclone, tornadoes, hurricanes, El Nino, La Nina and others. Some of the important ones are discussed in this section.

i. **Storm Fronts:** Storm Fronts can occur when air masses meet. An air mass is an enormous body of air with almost uniform humidity and temperature. The air masses are named according to regions, where they originate. Arctic and polar are cold; tropical and equatorial are warm; maritime is moist; continental is dry. When the air masses meet, a front is formed. This is the eye of the storm. Fronts can be cold or warm. Cold fronts create the most violent storm. Cold heavy air lifts the lighter warm, wet air causing an increase in humidity and formation of tall cloud, leading to storm. The vertical motion of the air is the key to the storm.

ii. **Cyclone:** A frontal cyclone is a large spinning weather system that occurs along turbulent fronts. Coriolis Effect causes the wind to veer right in the north and left in southern hemispheres, it also causes spinning motion. Frontal cyclones generally occur at the junction of the two fronts.

iii. **Tornadoes:** Most tornadoes occur in the US because of two opposing air masses, the cool air mass from Canada, and the warm, tropical air mass from south. In spring, when cold air mass still lingers, the warm tropical air mass from south moves in. The light warm air moves fast over the cold mass creating a spinning wind vortex, called tornado. Tornadoes wind can exceed 300 mph (480kph). One of the effects of global warming can be attributed to increased occurrences of tornadoes.

iv. **Hurricanes:** Hurricanes are tropical cyclones whose wind speeds exceed 73 mph (160km/h). Hurricanes are called typhoons in the western Pacific and cyclones in the Indian Ocean. Unlike frontal cyclone, hurricanes do not occur along the fronts of contrasting air masses. The terms "hurricane" and "typhoon" are regionally specific names for a strong "tropical cyclone". A tropical cyclone is the generic term for a non-frontal synoptic scale low-pressure system over tropical or sub-tropical waters with organized convection (*i.e.* thunderstorm activity) and definite cyclonic surface wind circulation.

v. **El Nino[3,6,7]:** 'El Nino' (Spanish; *the Christ Child*) was a term coined by Ecuadorian and Peruvian fishermen. It indicated the warm ocean phenomena, which occurs every 2 to 7 years around Christmas off the coast of Ecuador and Peru.

In the El Nino year, an easterly trade wind around the equator weakens. This allows warm water in the eastern Pacific, around Australia and Indonesia to flow eastwards. It flows towards Peru and Ecuador. The *thermocline* layer of water is the area of transition between the warmer surface waters and the colder water of the bottom[6].

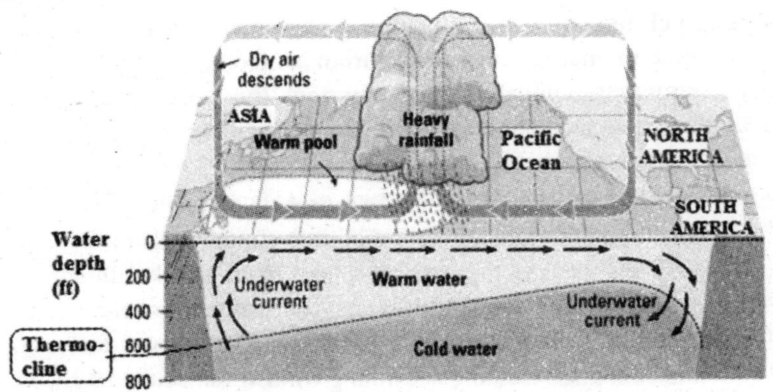

Fig. 2.4 EL Nino conditions and effects on climate across Pacific ocean, from Asia to North and south America

El Nino condition (fig.2.4) results from weakened trade winds in the eastern Pacific Ocean near Indonesia, allowing piled-up warm water to flow toward South America. This causes *upwelling* of warm water by about 500ft, pushes the *thermocline* (boundary between warm and cold water), to a lower level and results in disaster. The big circulatory loop moving the upper atmosphere from Australia & Indonesia to South America and then back along the surface is known as the *Walker circulation*[6].

El Nino events are linked with a change in atmospheric pressure known as Southern Oscillation (SO). The Southern Oscillation is the see-saw pattern of reversing surface air pressure between the eastern and western tropical Pacific.When the surface pressure is high in the eastern tropical Pacific it is low in the western tropical Pacific, and vice-versa. Because the ocean warming and pressure reversals are, for the most part, simultaneous, scientists call this phenomenon the *El Nino / Southern Oscillation or ENSO* for short.

The normal increase in water temperature is 1 to 2°C while in strong El Nino temperature can go up to 4-6°C. The temperature increase is at the higher side of the current prediction of global warming by 1.4 to 5.8°C over the next century. Thus *El Nino can be used as a working model of the effect of global warming on global weather.*

Impact of El Niño

In El Niño years, the rain area that is usually centered over Indonesia and the far western Pacific moves eastward into the central Pacific. The waves in the flow aloft are affected, causing unseasonable weather over many regions of the globe. The impacts of El Niño upon climate in temperate latitudes show up most clearly during winter time. For example, most El Niño winters are mild over western Canada and parts of the northern United States, and wet over the southern United States from Texas to Florida. El Niño affects temperate climates in other seasons as well. But even during wintertime, El Niño is only one of a number of factors that influence temperate climates.

El Nino in 1997-98 was one of the most powerful climate events of the century. According to UN estimates, massive fires, floods, frosts and droughts killed around 2100 people and caused property damages to the tune of $33 billions. Asia, Indonesia, and Japan were hit by famine and fire during the 1997-98 El Niño. Many economists believe that this was one of the reasons for the great Asian Stock Market Crash in 1997[2].

El Nino produces more heat, and together with the increase in global warming, it is expected to cause an intense disaster. If global warming can increase the sea water temperature in South America, the amount of increase is sufficient to kick in an El Nino, which may become a permanent feature.

According to NOAA (National Oceanic and Atmospheric Administration's recent report (July 2009), El Niño, the warming of the eastern and central Pacific Ocean, has developed and should persist throughout the winter, which may cause damaging storm in California and more storms in the South.

Ecosystem Destruction by El Nino

Coral reefs are sensitive ecosystems; they are a home to many plants and fish. The rise in sea temperature caused by El Nino and exposure to the Sun combine to destroy algae that protects the coral, which then bleaches white and dies. Destruction of coral from the effects of El Nino can be extensive. Recovery of the reefs may take a very long time.

vi. La Nina: La Nina is the wicked sister of El Nino. It is characterized by unusually cold ocean temperature in the Equatorial Pacific[2]. Therefore La Nina is the negative, or the cooling phase of El Niño. The effects are more intense from December through March, which is the same period as that of El Nino.

During La Nino, westbound intense trade wind drives more than normal amount of warm surface water westward, causing the rise of deep chilly water off the cost of Peru and Ecuador. This produces a cold tongue of water that extends from South America, all the way to Samoa (Fig.2.4). With heat (warm water) moving westward, the rising moisture in West Pacific forms an intense low pressure zone. This affects the monsoons off Southeast Asia, drenches part of Australia, and produces rains at far off places like southern Africa.

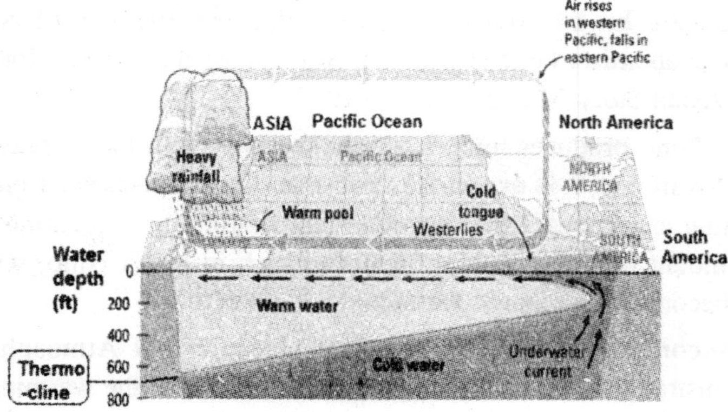

Fig. 2.5 La Nina conditions and effects across Pacific ocean from Asia to North & South America

La Nina formation is shown in fig.2.5(ref.7). Strong Pacific trade winds blow warm surface water westward. Cold water rises to the surface. Cooler air disrupts jet streams; northern jet stream loops into Alaska, Canada

During La Nina, the polar jet stream moving towards Canada, goes further south, thus pushing frigid air deep into USA, causing freezing winter in Northwest & Upper Mid-west, producing less rain in southeastern states and drought in southwest deserts. In 1998, La Nina hurricane was the deadliest, killing 9,000 in Central America. A probable cause of the tornadoes which swept the Southern US on 5th February, 2008 is La Nina, due to the cooling of tropical Pacific(2).

Natural disasters are likely to be more frequent and with greater intensities due to climate change resulting from global warming.

Summary

i. Weather is the current atmospheric conditions of temperature, pressure, humidity, and the resultant precipitation. The climate is the average of these parameters over a longer period. When the air becomes fully saturated with moisture, water vapor condenses to form tiny droplets of water or ice, depending on temperature, and precipitates as rain or ice respectively. Types of clouds, temperature and pressure determine the extent of rain or snow or thunderstorms.

ii. Atmosphere circulates from equator to poles in three circulation cells, which are known as the Hadley (30° N/S in latitude), Ferrel (30° -60° N/S) and Polar (60° -90° N/S).Global warming has extended Hadley's cell by 2 to 4.5 degrees in latitude, beyond the 1979 figures due to global warming.

iii. Due to Earth's rotation on its tilted axis, different parts of the Earth receive higher and lower levels of radiant energy. This creates four seasons, known as spring, summer, winter and autumn.

iv. Areas with similar climates are grouped together in climate regions, such as, Tropical (A), Dry (B), Exothermal or

Temperate, (C), Microthermal (D), and Polar (E), from equatorial region (A) upwards and Highland (H).

v. Disasters occur when there is a sudden change in weather resulting in, storms, cyclone, tornadoes, hurricanes and El Nino, & La Nina. Climate changes can result in more frequent disasters, with greater intensities, or changes in the characteristics of disasters.

REFERENCES

1. Roger G. Berry and Richard J.Chorley, *Atmosphere, Weather & Climate, Routledge*, London and New York, 7th edition, reprinted in 2002.
2. Michael Tennsen, *Global Warming*, Alpha Books, NY, USA, 2004.
3. David D. Kemp, *Exploring Environmental Issues –An integrated approach*, Routledge, London & New York, 2004.
4. Peter J Robinson & Ann Henderson-Sellers, *Contemporay Climatology*, second edition, Longman, 1999
5. *Hindustan Times*, Mumbai, India December 8, 2007, p. 19.
6. El Nino - and what is the Southern Oscillation Anyway, *Updated: January 22, 2003,* NASA Science Content Manager, *N*ASA website http://kids.earth.nasa.gov/archive/nino/normal.htm
7. El Niño and La Niña, *Weather's Sibling Rivalry in Dummies.*com http://www.dummies.com/how-to/content/el-nino-and-la-nina-weathers-sibling-rivalry.html?newComment=true#comments

4

Solar Beam and Global Heating

The temperature is a predominant factor in the atmospheric divide in five layers. Again the temperature is found to be responsible for three major air circulation systems creating six different climate zones extending from equator to poles in both north and south. The source of heat for atmospheric temperature rise is the Sun. The high energy solar radiations travel through space to the planet's atmosphere before striking the surface. The heat absorbed from the solar beam by the atmosphere and the Earth's surface leads to global warming. The characteristic features of solar beam and its heat transfer processes to the Earth and the atmosphere are to be discussed in this chapter.

The heat source for global heating is the Sun, which is a huge gas mass at an extremely high temperature. Solar radiation originates from nuclear fusion reaction and it takes less than ten minutes for the solar beam to strike the Earth's surface. Solar radiation consists of ionizing radiations and non-ionizing beams. The highly injurious ionizing radiations are filtered out by upper atmosphere, allowing only non-ionizing part to reach the Earth.The non-ionizing radiation consists of three bands, viz., ultra-violet (UV), visible and infra-red. The infra-red portion of the radiation generates most of the heat through absorption by some of the atmospheric gases, such as, carbon-dioxide.

The total quantity of incoming solar radiation depends on solar constant, solar irradiance and solar luminosity. The one-fourth of the solar constant, which is around 340 Watts per square metre reaches the Earth. However, of the total incoming radiation, around 30% is reflected back by surface, clouds, and particulate and the rest 70% is used in processes like melting ice, evaporating water, photosynthesis, absorption by cloud and atmospheric gases. Only 19% of solar energy is available for absorption by atmospheric greenhouse gases.

The solar beams striking the Earth cause the atmospheric temperature to raise. The variations in temperatures in different regions occur due to the different angles at which the solar beams strike the Earth's surface, The beam striking at 90° transmit the highest level of energy. Therefore the hottest places on earth are those areas where the rays of the sun arrive at right angles. Other locations, where the sun's rays hit at lesser angles, tend to be cooler.

Solar radiation is also responsible for evaporation, and consequently the hydrological cycle and the ocean currents. Solar radiation is responsible for photosynthesis, thus the carbon cycle. Hydrological and carbon cycles determine the quantities of heat (infra-red) absorbing water vapor and carbon dioxide present in the atmosphere and thus the quantum of heating. Among the factors responsible for climate change due to external forces, are *variations in the earth's orbit* around the Sun, and changes in *solar luminosity*.

Sun's Surface

The surface features of the Sun are normally classified into three regions. The three regions are known as the photosphere, chromosphere, and corona. The photosphere corresponds to the bright region normally visible to the naked eye. About 3,100 miles (5,000 km) above the photosphere is the chromosphere, from which short-lived, needle-like projections may extend upward for several thousands of kilometers. The corona is the outermost layer of the Sun; this region extends into the regions of the planets. Most of the surface features of the Sun lie within photosphere, though a few extend into the chromosphere and even the corona. The visible radiation (light) comes from photosphere[1].

Solar Radiation

Solar radiation originates from a *nuclear fusion reaction* in the sun, forming ionizing radiations (X-rays and gamma-rays) and non-ionizing radiations.

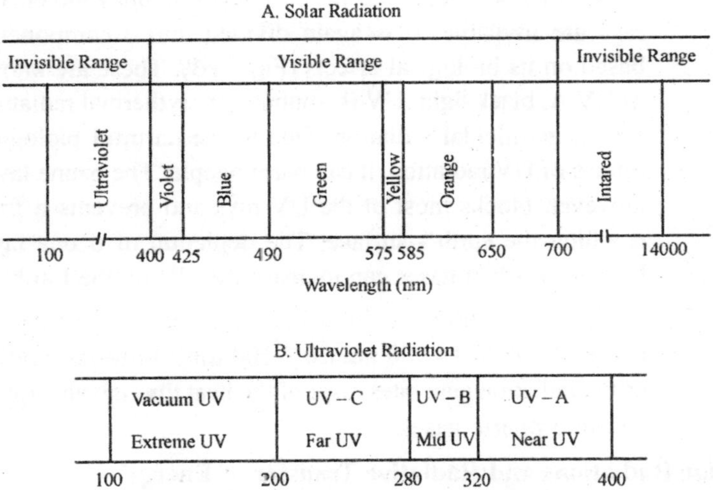

A. Solar Radiation

Invisible Range | Visible Range | Invisible Range

Ultraviolet | Violet | Blue | Green | Yellow | Orange | Intared

100 400 425 490 575 585 650 700 14000

Wavelength (nm)

B. Ultraviolet Radiation

Vacuum UV	UV – C	UV – B	UV – A
Extreme UV	Far UV	Mid UV	Near UV

100 200 280 320 400

Fig.3.1. Spectra of nonionizing solar radiation (A) and ultraviolet radiation (B) showing main radiation bands, other synonyms: UV-A, black light; UV-B, sunburn or erythermal radiation; UV-C (compiled from WHO 1979).

Fortunately, the highly injurious ionizing radiation does not penetrate the Earth's atmosphere. The non-ionizing part can permeate through the upper atmosphere.The spectral range of non-ionizing solar radiation (fig.3.1A) is commonly divided into three regions or bands on the basis of wavelength as follows:

1. Visible: The spectral band wavelength extends from 400 to 700nm. White light consists of seven colors in the beam, viz, violet, indigo, blue, green, yellow, orange and red. These are spread over a range of wavelength from violet at 400nm (0.4 ìm) to red at 700nm (0.7 ìm). Before violet on the short wavelength side is ultra violet (UV). After red are

infra-red (IR) rays, which have a longer wave length than red. Radiation in the visible range is not absorbed by the atmosphere.

2. Ultra-violet: UV rays are comprised of short wavelength solar radiation, in the range of 100-400 nm (0.001-0.4 μm). This wavelength range is lower than violet, on the extreme left of visible range, and hence known as ultra violet. UV rays are invisible. It is again divided into 3-components, based on its biological effects (Fig.3.1B). These are known as UV-A, black light; UV-B, sunburn or erythermal radiation; UV-C, germicidal radiation. Due to the harmful biological effects of UV radiation, it can harm people. The ozone layer, however, blocks most of the UV rays and prevents it from reaching the earth's surface. The depletion of ozone layer by fluorocarbon gases can increase the UV on the Earth.
3. Infra red: Infra-red is invisible, and covers the spectral range of 700 to 14,000 nm (,0.7-100 μm). Infra-red portion of the radiation generates most of the heat through absorption by atmospheric gases.

Solar Radiations and Radiative Transfer of Energy

Solar beams provide the energy required to propagate and sustain life on the Earth. The Sun is a huge gas mass, with core temperature of 5,500'K, at a distance of about 150 million kms and it takes approximately 10 minutes for sunlight to reach us. Solar beam provides the light and heat, by permeating through the atmospheric layers and finally striking the Earth's surface. The transfer of thermal energy or heat of the solar beams in the form of the electromagnetic wave can occur through conduction, convection and radiation. Major heat transfer occurs by radiation. In ideal conditions, a substance absorbs all the energy impinging on it from solar beam and becomes a 'black body'. The black body itself then becomes a radiation source, where incoming radiation balances the outgoing one.

The wavelength of propagation depends on the temperature of the emitting body, and the basic equation governing the relation between wavelength and Temperature is given by Planck's Law:

$$E_{\lambda *} = c_1 / \{\lambda^5 [\exp.(c_2 / \lambda T) - 1]\}$$

Where $E\lambda_*$ = amount of energy (W m^{-2} μm^{-1}) emitted at wavelength λ (µm) by a body at a temperature of T in degree Kelvin. c_1 and c_2 are constants. By plotting wavelength against energy flux, "bell-shaped" curve results, showing energy distribution over a range of wavelength. However, the energy of solar radiation over the entire range of wavelength *i.e.*, 0.001 to 100 µm follows the Gaussian distribution pattern and thus sharply centered on the wavelength band of 0.2-2 µm

By integration of the area under the Planck's curve, the total energy (E) emitted by a black body is found to be directly proportional to the fourth power of the absolute temperature (T). The resultant equation is known as Stefan – Boltzman Law is as follows:

$$E = \sigma T^4$$

Where σ = Stefan – Boltzman constant =5.67 × 10^{-8} W m^{-2} K^{-4}

The total solar input to space assuming a temperature of 5760 K for the sun becomes 3.84 × 10^{26} Watts. However, only a small fraction is intercepted by the Earth, primarily due to the fact that the energy received is inversely proportional to the square of the solar distance (150 million Km).

By differentiation of Planck's equation, the equation relating wave length and temperature is obtained, which is as follows:

$$\lambda max = 2897 / T$$

The above equation is known as Wien's Law. According to Wien's equation the maximum energy wavelength is inversely related to temperature. The equation indicates the maximum emission wavelength for the Sun and the Earth are 0.50 and 11.4µm respectively. The intense solar radiation is mainly in shortwave range of 0.2 to 4.0 µm, with maximum emission at 0.5 µm. The weaker terrestrial radiation is in the range of 4 to 100 micron with a peak at about 10 µm.

Of the 99% of solar energy emitted is in the range from ultra violet to infra red. According to theoretical calculations, based on

Planck's law, the shares of ultraviolet, visible and infra-red are 9%, 45%, and 46% respectively[2].

The 46% of the solar radiation in the infra-red range is available for absorption by greenhouse gases. *Anthropogenic greenhouse gases constitute the main source for providing extra heat to the Earth's atmosphere than that provided by omnipresent greenhouse gases formed by natural processes.*

Climate Model

Climate model simulates the quantitative interactions of the atmosphere, ocean, land surface and ice, in terms of temperature, pressure and precipitation. The climate model is used to study the dynamics of the climate system and to project future climate The recent application of climate model includes projection of temperature changes due to increases in greenhouse gases in the atmosphere. All climate models take into account incoming solar energy (short wave radiation, mainly,visible and near infra-red) and the outgoing energy(long waves, far infra-red)radiation from earth. The imbalance between the two results in a change in temperature

Effective temperature of the planet by solar emission

Climate model

By denoting the solar radiative flux at the top of the planet's atmosphere by S_0 (for solar constant), the Earth's average albedo by a, and planet's radius as R the area of that disk is πR^2

Heat absorbed by planet = $(1 - a) \pi R^2 S$

The total heat radiated from the planet is equal to the energy flux implied by its temperature, T_e (from the Stefan-Boltzmann law) times the entire surface of the planet or:

Heat radiated from planet = $(4 \pi R^2) \sigma T^4$

Radiative balance can therefore be written as follows:

$(4 \pi R^2) \sigma T_e^4 = (1 - a) \pi R^2 S_0$

This is a very simple model of the radiative equilibrium of the Earth

Solving this equation for temperature we obtain:

$T_e = [(1-a)S_o / 4\sigma]^{1/4}$

The equation above yields an average earth temperature of 288°K (15 °C; 59 °F)[(2a)]. The above equation represents the effective *radiative* temperature of the Earth (including the clouds and atmosphere).The use of effective emissivity and albedo account for the greenhouse effect. With this temperature the Earth radiation will be centered on a wavelength of about 11 μm, well within the range of infra-red (IR) radiation.

The Earth and other planets are not perfect black bodies, as they do not absorb all the incoming solar radiation but reflect part of it back to space. *The ratio between the reflected and the incoming energies is termed the planetary albedo.*

Solar Constant[1,2]

The solar constant is the average amount of incoming solar radiation per unit area, falling on the outer surface of Earth's atmosphere, in a plane perpendicular to the rays. Solar constant is approximately 1,367 watts per square meter. As the earth-based measurements of the quantity are not accurate, due to variations in the earth's atmosphere, scientists rely on satellites to make these measurements. Although referred to as the solar constant, this quantity has actually been found to vary since careful measurements were made in 1978. In 1980, a satellite-based measurement yielded the value of 1,368.2 watts per square meter. Over the next few years, the value was found to decrease by about 0.04% per year. Such variations have now been linked to several physical processes known to occur in the Sun's interior.

Solar Irradiance[1, 2, 3]

Solar irradiance, or insolation, is the amount of sunlight which reaches the Earth. From the Earth, it is only possible to observe the radiant energy emitted by the Sun in the direction of our planet. Irradiance is evaluated through measurements of optical brightness, total radiation, or radiation in various frequencies. Both the radiation reaching the upper atmosphere and that to the Earth's surface makes up the total irradiance. The gases in the atmosphere absorb some solar radiation at different wavelengths. The clouds and dust also affect the solar irradiance. The average incoming solar radiation or the solar irradiance, taking into account the half of the planet not receiving any solar radiation at all, is one fourth the solar constant or ~342 W/m². At any given location and time, the amount received at the surface depends on the state of the atmosphere and the latitude.

Solar Luminosity

The total radiant energy emitted from the Sun in all directions is a quantity known as solar luminosity. The luminosity of the Sun has been estimated to be 3.8478 x 10^{26} watts. Some scientists believe *that long-term variations in the solar luminosity may be a better correlation to environmental conditions on Earth than solar irradiance, including global warming.* Variations in solar luminosity can occur due to stellar rotation, convection and magnetism.

The measurements of total solar irradiance, taking into account the solar flux contributions over all wavelengths, provide a more accurate value than total luminosity because of the short time variations of certain regions of solar spectrum. Short-term variations in solar irradiation vary significantly with the position of the observer particularly more on ground based measurements. However satellite based measurements give accuracy within few parts per million, allowing scientists to acquire a better understanding of variations in the total solar irradiance.

Variations in solar irradiance have been attributed to the solar phenomena, such as, oscillation, granulation, sunspots, faculae, and solar cycle.

Oscillations, arise from the action of resonant waves trapped in the Sun's interior. At a given time, there are tens of millions of frequencies represented by the resonant waves, but only certain oscillations contribute to variations in the solar irradiance, lasting for few minutes (about 5 minutes).

Granulation, produces solar irradiance variations lasting about 10 minutes. It is closely related to the convective energy flow in the outer part of the Sun's interior. To the observer on Earth, the surface of the Sun appears to be made up of finely divided regions known as granules, separated by dark regions. Each of these granules makes its appearance for about 10 minutes and then disappears. These granules are believed to be the centers of rising convection cells.

Sunspots give rise to variations that may last for several days, and sometimes as long as 200 days. They actually correspond to regions of intense magnetic activity where the solar atmosphere is slightly cooler than the surroundings. Sunspots appear as dark regions on the Sun's surface to observers on Earth. The reduction in the total solar irradiance has been attributed both to the presence of these sunspots and the formation of newly formed active regions associated with large sunspots, or with rapidly evolving, complex sunspots.

Faculae, producing variations that may last for tens of days, are bright regions in the photosphere where high-temperature interior regions of the Sun radiate energy. They tend to congregate in bright regions near sunspots, less dense than the surroundings, forming solar active tube-like regions defined by magnetic field lines. The radiation from hotter layers below the photosphere can leak through the walls of the faculae, thus producing an atmosphere that appears hotter, and brighter, than others.

Solar cycle is responsible for variations in the solar irradiance that have a period of about 11 years. This 11-year activity cycle of sunspot frequency is actually half of a 22-year magnetic cycle, which arises from the reversal of the poles of the Sun's magnetic field. From one activity cycle to the next, the north magnetic pole becomes the south magnetic pole, and vice versa. *Solar luminosity has been found to achieve a maximum value at the very time that sunspot activity is highest during the 11-year sunspot cycle.* Surprisingly, the Sun's rotation, with a rotational period of about 27 days, does not give rise to significant variations in the total solar irradiance. This is because its effects are overridden by the contributions of sunspots and faculae.

Solar Irradiance and Global Warming

The long-term solar irradiance variations may contribute to global warming over decades or hundreds of years. In recent times, there has been speculation that changes in total solar irradiation have amplified the greenhouse effects, *i.e.*, the retention of solar radiation and gradual warming of the earth's atmosphere. Some of these changes, particularly small shifts in the length of the activity cycle, seem to correlate rather closely with climatic conditions in pre- and post industrial times. However any major effect of variations in solar irradiance on global warming over the past 150 years remains highly controversial.

Atmospheric Effects on Incoming Solar Radiation

Solar radiation is partially depleted and attenuated as it traverses the atmospheric layers, preventing a substantial portion of it from reaching the Earth's surface. This phenomenon is due to absorption, scattering, and reflection in the upper atmosphere (stratosphere), with its thin layer of ozone, and the lower atmosphere (troposphere), within which cloud formations occur and weather conditions manifest themselves.

The *stratospheric ozone layer* has a strong absorption affinity for solar UVR, depending on wavelength. Depletion of the protective ozone layer beyond the critical level by certain atmospheric pollutants (fluorocarbons and nitrogen oxides) shall promote the transmission of highly injurious UVR.

The *troposphere* is an attenuating medium. Direct solar radiation received by the earth's atmosphere and surface is modified by atmospheric scatterings. The solar radiation is reflected and scattered primarily by clouds (moisture and ice particles), particulate matter (dust, smoke, haze, and smog), and various gases. The *diffused solar radiation* received by earth's surface is that has been modified by atmospheric scattering. The two major processes involved in troposphere scattering are determined by the size of the molecules and particles and are known as *selective scattering* and *nonselective scattering.*

The amount of scattering that takes place is dependent on two factors: wavelength of the incoming radiation and the size of the scattering particle or gas molecule. The degree of scattering decreases with increasing wavelength in the following order: UV-B > UV-A > violet > blue > green > yellow > orange > red > infrared. The presence of a large number of particles with a size of about 0.5 microns in atmosphere, results in shorter wavelengths being preferentially scattered. Effects of scattering include, reduction of the amount of incoming radiation reaching the Earth's surface and a significant proportion of scattered shortwave solar radiation is redirected back to space. Without scattering in our atmosphere, the daylight sky would be black.

Selective scattering is caused by smoke, fumes, haze, and gas molecules that are either equal to or smaller than the wavelength of incident radiation. Selective scattering is severe when atmosphere is extensively polluted. Selective scattering in the blue region of the spectrum by smaller molecules under clear-sky conditions accounts for the blue sky. The color of the sky is also determined by the length of the atmospheric path traversed by sunlight.

Nonselective scattering is caused by dust, fog, and clouds with particle sizes more than 10 times the wavelength of the incident radiation. As scattering in this case is not wavelength-dependent, it is equal for all wavelengths, *i.e.*, to all colors. This factor causes the clouds to appear white.

Absorption of Radiation by the Atmosphere and Heating Effect

Absorption is defined as a process in which solar radiation is retained by a substance and converted into *heat energy*. Visible range is not absorbed by atmospheric gases. Spectral irradiance in visible wavelength varies from 1.7W/m^2/nm at the top of the atmosphere, to 0.4 W/m^2/nm at sea level where it is not absorbed by atmospheric gases.Only Infrared spectrum with irradiance of 0.7 W/m^2/nm is absorbed by H_2O. IR spectrum with irradiance of 0.1 W/m^2/nm is absorbed by CO_2 and H_2O. at sea level. The radiation in the wavelength range of approx. 2000 to 2500 nm in the infra red is totally absorbed by CO_2 at the sea level.

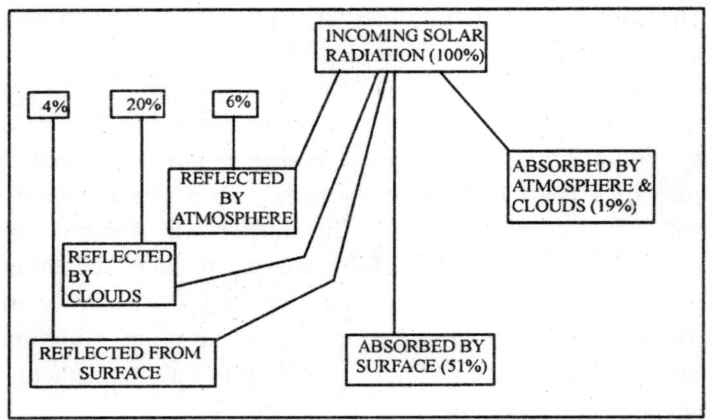

3.2 Solar radiation, surface & atmospheric reflection & absorption

Reflectivity or Albedo

As in the atmosphere, some of the radiation received at the Earth's surface is redirected back to space by reflection. The *reflectivity or albedo* of the Earth's surface varies with the type of material that covers it. For example, fresh snow can reflect up to 95% of the solar beams that reaches it surface. Some other surface reflectivities include, dry sand 35 to 45%, broadleaf deciduous forest 5 to 10%, needle leaf coniferous forest 10 to 20%, and grass type vegetation 15 to 25%.

The albedo at two levels, viz., the Earth's surface and the top of the atmosphere, are of great importance. The top of the atmosphere one is known as *planetary albedo* helps in understanding of large scale climate processes. The *surface albedo* depends largely on the nature of the surface and has significant effect on local climate. The Earth's average *albedo*, reflectance from both the atmosphere (planetary albedo) and the surface (surface albedo), is about 30%. The other albedo, the *cloud albedo,* plays an important role in the atmospheric heating and thus has influence on local climate.

Distribution of Incoming Solar Energy (fig.3.2)

Of all the total incoming sunlight that passes through the atmosphere annually, *51% is available at the Earth's surface* to do works, such as,

(i) heat the Earth's surface at lower atmosphere,

(ii) melt ice and evaporate water, and

(iii) run photosynthesis in plants.

These are essential functions to maintain natural ecological cycles, such as hydrological and carbon cycles.

The *remaining 49% of solar beams is distributed* as follows:

i. 4% is reflected back to space by the Earth's surface,

ii. 26% is scattered or reflected to space by clouds and atmospheric particles and

iii. 19% is absorbed by atmospheric gases, particles, and clouds.

Average incoming solar radiation is around 342 watts per sq.meter. The distribution of solar energy in various functions is as follows:

i. 51% absorbed by surface amounts to 174.4 W per sq.meter

ii 19% absorbed by atmosphere equals to 64.98 W/sq.meter.

iii. 30% reflected beams account for rest 102.6 W/sq.meter Reflections include by atmosphere as 6% or 20.52W/ sq.meter,by cloud as 20% or 68.40 W/sq.meter and by surface as 4% or 13,68 W/sq.meter.

The 19% of solar beam absorbed by atmospheric gases, mainly carbon dioxide, plays the key role in global atmospheric warming. Deforestation leads to less absorption of solar beam required by plants for photosynthesis.Deforestation and higher rates of carbon dioxide emission shall cause accelerated heating of the atmosphere by incoming solar beams.

If the 51% of solar energy amounting to 174.4 W per sq.meter absorbed by the Earth's surface can be utilized for power generation, then there would be no need to burn fossil fuel to generate energy. In such circumstances, there would not be any enhanced rate of global warming from anthropogenic emissions. The intensity of sunlight at ground level varies with latitude, geographic location, season, cloud coverage, atmospheric pollution, elevation above sea level, and solar altitude. Efforts to harness the solar energy in high intensity areas, such as, deserts for power generation is already on the anvil (chapter 12).

External Forcing and Climate Change

Radiative Forcing

The energy budget of the Earth is considered to be in equilibrium, maintains a balance between incoming solar radiation and outgoing terrestrial radiation over the long term. Radiative forcing is defined as the change in the balance between solar radiation entering the atmosphere and the Earth's radiation going out.

The heat balance of the atmosphere is liable to change due to changing heat absorption capacities (*e.g.*, greenhouse gases) and heat reflection abilities (*e.g.*, albedo) of the materials.The imbalance causes changes in global temperatures. On average, a positive radiative forcing tends to warm the surface of the Earth while negative forcing tends to cool the surface.

The concept of radiative forcing is useful because a linear relationship has been determined between the global mean equilibrium surface temperature changes and the amount of RF. The radiation balance can be altered by factors such as intensity of solar energy (solar irradiance), reflection by clouds or gases (albedo), absorption by various gases or surfaces (greenhouse gases), and emission of heat by various materials (volcanic eruptions). Any such alteration is a radiative forcing, can cause a new balance to be reached.

Climate responds to several types of external forcing, such as radiative forcing due to changes in atmospheric composition (mainly greenhouse gas concentration), variations in Earth's orbit (orbital forcing), changes in solar luminosity, and volcanic eruptions[4]. The contributions of the some major components of radiative forcing as assessed by IPCC in 2005 are shown in fig.3.3 (ref.5).

Orbital Forcing

The 23.5° tilt of the Earth's axis affects the angle of incidence of solar radiation on the earth's surface and causes seasonal and latitudinal variations in day length. The slow changes in the tilt and the shape of the orbit (Milankovitch cycle) cause changes in climate, a process known as *orbital forcing*. These orbital changes alter the total amount of sunlight reaching the Earth by up to 25% at mid-latitudes (from 400 to 500 Wm^{-2} at latitudes of 60 degrees).

The net radiative forcing due to anthrpogenic components amounts to around (+)1.82W/mm.sq(Fig.3.3).Land use change (including urbanization, deforestation, reforestation, desertification, etc) can have significant effects on radiative forcing (and the climate) at the local level by changing the reflectivity of the land surface (or albedo).

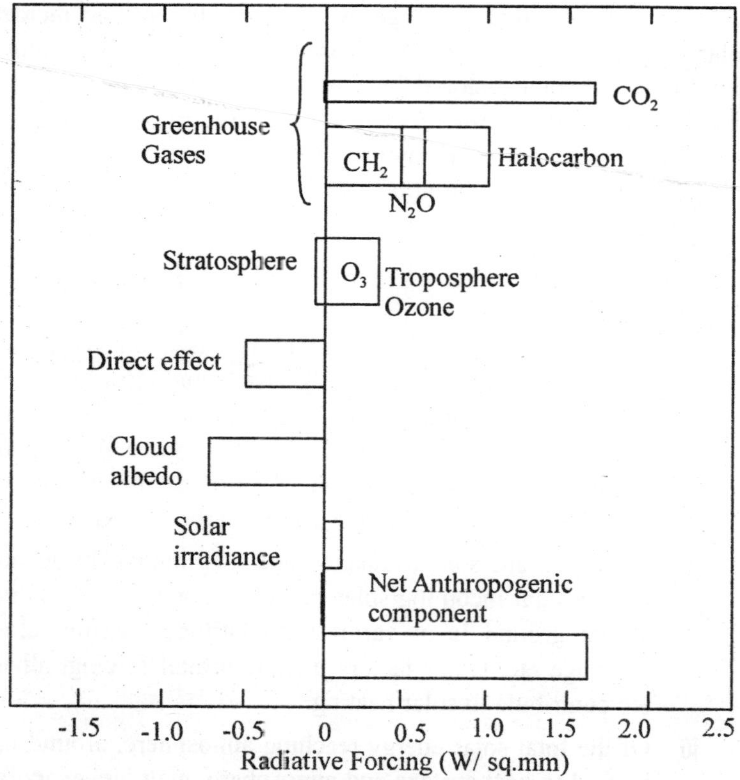

Fig. 3.3 Major Radiative Forcing Componen and net Anthropogenic Componnt

Solar luminosity

Solar radiation is unevenly distributed throughout the world because of such variables as solar altitude (associated with latitude and season) and atmospheric conditions (cloud coverage & degree of pollution). At high altitudes, the intensity of UVR is significantly higher than at sea level. The spectral distribution of solar energy at sea level is roughly 3, 44, and 53% in the UV, visible, and infrared regions, respectively.

The most favorable belt (15-35° N) encompasses many nations in northern Africa and southern parts of Asia. This belt recieves over 3000 hours per year out of 4380 (365x12) hours total sunshine with the limited cloud coverage. More than 90% of the incident solar radiation comes as direct radiation. The moderately favorable belt (0-15° N), or equatorial belt, has high atmospheric humidity and cloudiness that tend to increase the proportion of scattered radiation. The global solar intensity is almost uniform throughout the year as the seasonal variations are only slight.

Summary

i. The Earth's source of energy is the Sun. High energy solar radiation originates from a *nuclear fusion reaction* in the sun. Three parts of solar radiation, based on wavelength are termed as ultra-violet (UV), visible and infra-red. UV is filtered out by stratospheric ozone layer. The visible and infra-red portions reach the Earth's atmosphere. Atmosphere is heated through absorption of infra-red portion of the solar beam by some gases, such as, carbon dioxide.

ii. Solar constant, solar irradiance, and solar luminosity account for the total incoming solar radiation. Intensity of sunlight at the ground level varies with latitude, season, cloud coverage etc. Other factors include orbital forcing, albedo etc contribute to solar heating.

iii. Of the total solar energy reaching atmosphere, around half is used to heat surface and atmosphere, melt ice, evaporate water, and run photosynthesis, while from the rest half a major portion is reflected back by surface. Only 19% of solar radiation is available for absorption by atmospheric gases. Deforestation leads to more solar energy and carbon dioxide available to atmosphere and thus causing temperature to rise globally. Additional quantities of carbon dioxide emission add to more heating of the atmosphere.

iv. If 51% solar energy absorbed by the Earth's surface can be used to generate power, the global warming due to anthropogenic emission can be eliminated.

v. Climate responds to several types of external forcing, such as radiative forcing due to changes in atmospheric composition (mainly greenhouse gas concentration), variations in Earth's orbit (orbital forcing), changes in solar luminosity, and volcanic eruptions.

REFERENCE

1. R.G. Berry and R.J.Chorley, *Atmosphere, Weather, & Climate*, Chapter2, Solar Radiation and Global Energy Budget, p20-50, Routledge, NY, 2002.
2. Peter J Robinson & Ann Henderson-sellers, *Contemporary Climatology*. Chapter 2, The Earth's Radiation Budget, p17 -38, Longman, 2nd edition, 1999.

2a. The Institute for Global Environmental Strategies http://www. strategies.org,NASALG-2001-4-027-GSFC,ttp://eospso.gsfc. nasa.gov/ftp_docs/ lithographs/CERES_litho.pdf

3. Total Solar Irradiance - Global warming, Measuring solar irradiance http://science.jrank.org/pages/6875/Total-Solar-Irradiance.html#ixzz0biL6PNLn
4. Hegerl, Gabriele C.; *et al.* (2007). "Understanding and Attributing Climate Change" (PDF). *Climate Change 2007: The Physical Science Basis. Contribution of Working Group I to the Fourth Assessment Report of the Intergovernmental Panel on Climate Change*. IPCC. http:// ipcc-wg1.ucar.edu /wg1/ Report/ AR4WG1_Print_Ch09.pdf
5. IPCC,2007: Summary for Policymakers; In 'Climate Change 2007;The Physical Source Basis; Contribution of the Working Group 1 to the 4th Assessment Report of the IPCC,Cambridge University Press, Cambridge, UK, http://www.ipcc.ch/pdf/assessment-report/ar4/wg1/ ar4-wg1-spm.pdf

5

Global Warming and Climate Change

The atmosphere absorbs part of the solar beam while reflecting rest back into space. The capacity to absorb solar energy by the atmospheric gases determines the average temperature of the atmosphere. Some of the gases, such as carbon dioxide can absorb infra red or heat energy portion of solar beam and by this process can increase the atmospheric temperature. Although present in small quantity, the variation in the amount of carbon dioxide can cause differences in the atmospheric temperature.

Scientist throughout the last two centuries found enough evidence that some of the gases in our atmosphere allow sunlight but absorb the solar heat. Eight decades ago, it was George S. Callendar, who attributed the rise in global temperature due to carbon dioxide formed by burning of fossil fuels. However, several decades afterwards, it was Charles David Keeling who finally established the correlation of global warming with the increase in carbon-dioxide in the atmosphere. He accomplished this through collection of remarkably consistent and accurate data on the continued increase in carbon dioxide, a major greenhouse gas, for a prolonged period of time. Temperature being an integral part of climate, global warming can lead to climate changes.

The Earth's climate changes in response to external forces, including *greenhouse gases*, *variations in its orbit* around the Sun, changes in *solar luminosity*, and volcanic *eruptions*. There are also Earth's own variations in temperatures, for which UNFCCC uses the term *climate variability*. However, the climate change due to excessive emissions from industries, known as, *anthropogenic climate change* depends mainly on the extent of man-made greenhouse gas emissions to the atmosphere. The major source of greenhouse gas is fossil fuel burning, while the forest remains a major sink.

In a 2001 report, an intergovernmental panel of experts on climate change (the IPCC) concluded that the net effect of feedbacks from global warming "is always to increase projected atmospheric CO_2 concentrations". The mechanisms other than the greenhouse effect, such as, volcanic activity and changes in the sun's brightness cannot explain the temperature increase of the past 100 years.

Global Warming and Heat Absorbing Atmospheric Gases

George Callendar (1936)[1] based on the data on temperatures taken at weather stations and on shipboards for decades, concluded that the world was gradually warming. Callendar attributed *the rise in global temperature due to carbon dioxide formed by burning of fossil fuels*. He was much ahead of his time.

Several decades after George Callendar's observation of global temperature rise due to carbon dioxide, it was Charles David Keeling who finally established the correlation of global warming with the increase in carbon-dioxide in the atmosphere. He accomplished this with his collection of remarkably consistent data for a prolonged period of time.

Gas Absorbing Capacity of Ocean

Scientists felt that the ocean being a sink for carbon dioxide, would absorb the extra carbon dioxide. However, Scripps Institute of Oceanography found that the *carbon dioxide absorption capacity of sea water is limited*[2]. In addition to salt, the sea water is a complex mixture of chemicals. These chemicals create a peculiar buffering mechanism that stabilizes the acidity of sea water. The mechanism had been known for decades, but Reveille's works[2] indicate that sea water surface layer has a limited capability of absorbing carbon dioxide, barely one-tenth of the calculated buffering capacity.

Gas Absorbing Capacity of Forests

Deforestation has led to decrease in the capacities for absorbing greenhouse gases by the forests. The increasing quantities of carbon dioxide formation due to rapid industrialization and growing energy requirements coupled with the decreasing absorption capacities of the ocean and forest have resulted in continuous increase in carbon dioxide concentration in the atmosphere.

Charles David Keeling and Keeling Curves[3,4]

The data on the increase in quantities of carbon dioxide annually in the atmosphere over several decades were collected for the first time by Charles David Keeling (1928-2005), an US Scientist, who is considered as the father of global warming issues. The dedicated scientist spent decades in collecting these data at Mauna Loa observatory at Hawaii, before making the famous *Keeling curve*. The Mauna Loa record shows a continuous increase in the annual concentration of carbon dioxide, an important greenhouse gas, thus establishing the role of human contribution to global climate change.

Keeling's pioneering work is remarkable because of the following reasons:

i. **Developed a Device to Measure Carbon dioxide Accurately in the Atmosphere at Parts per million (ppm) Levels:** The current figure for carbon dioxide content in the atmosphere is approximately 0.038%, which in absolute quantity is 0.00038 or 380 parts per million. Hence it was absolutely essential to have measurement accuracy at ppm level and beyond, if the co-relation between increasing carbon dioxide and global warming was to be established. Average annual increase in Keeling's measurement in 1959-60 periods was 0.92 ppm, less than 1 ppm and thus conforms to measurement accuracy.

In 1958 C.D. Keeling, while working at the University of California in San Diego introduced a new technique for the accurate measurement of atmospheric CO_2. Keeling used cryogenic condensation of air samples, followed by NDIR spectroscopic analysis against a reference gas, using manometric calibration. Subsequently, this technique was adopted as an analytical standard for CO_2 determination throughout the world, including the World Meteorological Association (WMO). Keeling's data on atmospheric concentration of carbon dioxide at Mauna Loa, from 1959 to 1995 are indicated in tab.4.1

ii. **Collected Data at different Locations:** CO_2 measuring stations are distributed across the globe. Most, however, are located in coastal or island areas in order to obtain air without contamination from vegetation, organisms and industrial activity, *i.e.* to establish the so-called background level of CO_2. In considering such measurements, account should be taken of the established fact that land-derived air flowing seawards loses about 10 ppm of its carbon dioxide to dissolution in the oceans, and even more in colder waters (Henry's Law). Mauna Loa has an active volcano. However, Keeling and his team made measurements on the incoming ocean breeze and above the thermal inversion layer to minimize local contamination from volcanic vents. In addition, the data are normalized to negate any influence from local contamination.

Measurements at many other isolated sites have confirmed the long-term trend shown by the Keeling Curve, though no sites have a record as long as Mauna Loa. Due in part to the significance of Keeling's findings the NOAA began monitoring CO_2 levels worldwide in the 1970s. Today, CO_2 levels are monitored at about 100 sites around the globe.

Table 4.1: Atmospheric Concentration of CO_2 at Mauna Loa, Hawaii

Year	Jan. / Sept.	Feb. / Oct.	March / Nov.	April / Dec.	May / Annual	June	July	Aug.
	CO_2 Concentration in parts per million (ppm)							
1959	315.42	316.31	316.5	317.56	318.13	318	316.39	314.65
	313. 68	313.18	314.66	315.43	315.83			
1965	319.27	320.28	320.73	321.97	322	321.71	321.05	318.71
	317.66	317.14	318.7	319.25	319.87			
1970	324.89	325.82	326.77	327.97	327.9	327.5	326.18	324.53
	322.93	322.9	323.85	324.96	325.52			
1975	330.23	331.25	331.87	333.14	333.8	333.43	331.73	329.9
	328.4	328.17	329.32	330.59	330.99			
1980	337.84	338.19	339.91	340.6	341.29	341	339.39	337.43
	335.72	335.84	336.93	338.04	338.52			
1985	344.79	345.82	347.25	348.17	348.74	348.07	346.38	344.52
	342.92	342.62	344.06	345.38	345.73			
1985	344.79	345.82	347.25	348.17	348.74	348.07	346.39	337.43
	342.92	342.62	344.06	345.38	345.73			
1990	353.5	354.55	355.23	356.04	357	356.07	354.67	352.76
	350.82	351.04	352.7	354.07	354.04			
1995	359.98	361.03	361.66	363.48	363.82	363.3	361.93	359.49
	358.08	.57.77	359.57	360.69	360.9			

Ref. "Atmospheric CO_2 Concentration from Mauna Loa, Hawaii", C.D. Keeling and T.P. Whorff, Scripps Institution of, Oceanography, University of California

iii. **Collected Data for Every month of the Year:** Keeling collected data on carbon dioxide variation for every month of the year to establish seasonal variation in emission and absorption by plants. The variations in monthly data are related to seasonal carbon dioxide levels in relation to global plant cycle. Carbon dioxide was high in winter, falling in spring, low in summer, and rising again in fall. In spring and summer, the plants require carbon dioxide for growing leaves, flowers, and fruits. Tab.4.1 shows Keeling's data of 5-years interval, but measured and recorded every year.

iv. **Collected Data over a Long Period:** The measurements of carbon dioxide concentrations in the atmosphere by Keeling and his team were made over a long period of 36years (1959-1995).

In 1959, the highest CO_2 level was in the month of May at 318.13 ppmv an increase of 2.71 ppmv from January figure of 315.42ppmv (tab.4.1). At the end of the year, in December, the CO_2 content in the atmosphere was 315.43ppmv, showing an almost complete carbon balance in that year and place (see fig. C-cycle in next chapter).

The carbon dioxide data showed an increase in average value from 315.83 ppmv in 1959 to 324.52 ppmv in 1970,followed by 338.52 ppmv (1980) and 354.04 ppmv (1990)(tab.4.1).*The results indicate a increase in carbon dioxide in every decade and at higher rates than the preceding decade. The increase in the period, 1959 to 1970 is 9.69 ppmv, in the following decade, 1970-1980, is 10.00ppmv, and in the next decade.1980-1990 as 15.52 ppmv*

Keeling Curves

The results of the extensive data collection were graphically presented in the famous Keeling curves (Fig.4.2). Beginning in 1955, Keeling collected air samples to measure their carbon dioxide content. His measurements over the decades that followed showed that carbon dioxide levels were steadily rising — a finding that shattered the conventional wisdom that Earth could soak up rising fossil fuel emissions without harm.

The measurements at Mauna Loa, extended to 2006, (Fig.4.2) show a steady increase in average CO_2 concentration in atmosphere from about 315 parts per million by volume (ppmv) in 1958 to over 380 ppmv by the year 2006. That is 65 ppmv in last 40 years, 15.96 ppmv over the decade 1990-2000, and 10 ppmv over six years' period from 2000-2006. As indicated earlier the more recent data also indicates the same trend of increasing quantities of CO_2 in the atmosphere per year at a higher rate every year than the preceding one.

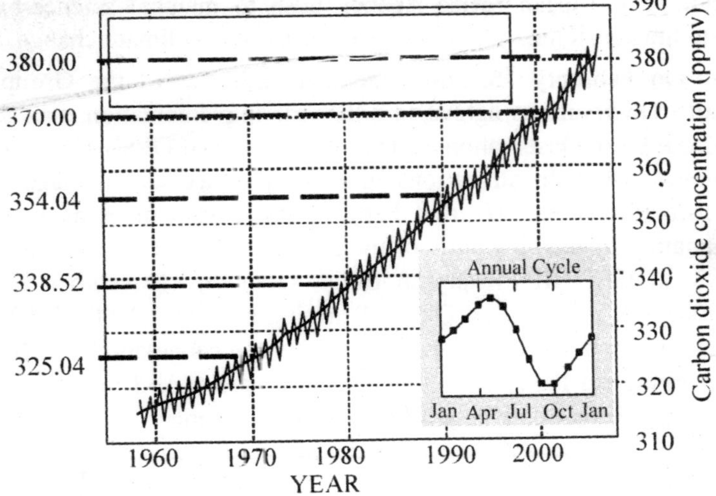

Fig. 4.2. The Keeling Curve; Atmospheric Carbon dioxide (ppmv) concentration in five decades.

This increase in atmospheric CO_2 at an accelerated rate in recent year is considered to be largely due to increasing quantities of combustion of fossil fuels, to cater to the growing energy need. This is supported by measurements of carbon dioxide concentration in ancient 'air bubbles' trapped in polar 'ice cores', which show that mean atmospheric CO_2 concentration was between 275 and 285 ppmv for several thousand years but started rising sharply at the beginning of the nineteenth century. The carbon dioxide absorbs most of the infra-red energy thus making significant contribution in heating up the atmosphere. The atmosphere around the globe shall absorb more heat with the increase in concentration of carbon dioxide with every passing year. The 'global warming' would occur at an accelerated rate in conformity with the increased volume of CO_2 accumulating in the atmosphere every year

Monitoring Weather

Monitoring our planet's weather, oceans and land masses, is of paramount importance to understand, forecast, and possibly manage Earth's ecological goods and services in the face of global warming. This is being done by satellites dedicated to provide such information. Both the Global Earth Observation System of Systems (GEOSS) and the European Space Agency push to marshal science-based satellite constellations for a concerted focus on climate change.

On February 16, 2005, member countries of the Group on Earth Observations agreed to a 10-year implementation plan for a Global Earth Observation System of Systems (GEOSS). The GEOSS project will help all nations involved produce and manage their information in a way that benefits the environment as well as humanity by taking a pulse of the planet. EPA-ORD represented the U.S. at GEO III Plenary in Bonn, Germany (November 27-29, 2006). UN agencies on Global Warming include, UNFCCC(United Nations Framework Convention on Climate Change)and IPCC (Inter-governmental Panel on Climate Change) are tasked to co-ordinate the activities around the world on global warming.

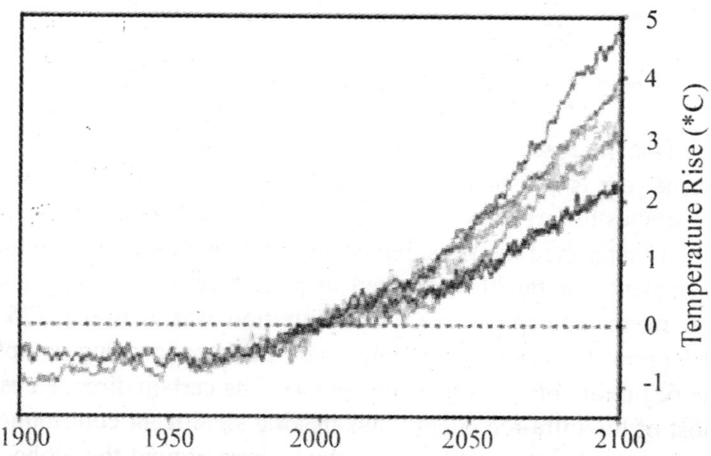

Fig. 4.3 Global Warming Projection (projection by CCSR/ NIES, CCCma, CSIRO, Hadley Centre, GFDL MPIM, NCAR PCM, & NCAR CSM)

Global Warming Projection

Global warming has been observed to increase in the average of the Earth's near-surface air temperature, which rose by 0.74 °C (1.3 °F) during the last century. Natural cycles of formation and absorption of carbon dioxide leads to seasonal climate changes by absorbing different quantum of heat by the amount of carbon dioxide available in the atmosphere at that period. However, the enhanced concentration of greenhouse gases (GHGs) from anthropogenic sources in atmosphere can induce a global average temperature rise of over 2°C as early as 2035. Various agencies have shown an increase in temperature from 2°C to 5°C in 2100 AD (fig.4.3). The increase in global temperature at these levels shall have great impact on the climate across the world. In the longer term, there would be more than a 50% chance that the temperature rise would exceed 5°C. This rise would be equivalent to the change in average temperatures from the last ice age to today. To avoid the worst effects of global warming the mean global temperature rise must be kept below 2° C (3.6° F). In a decade or two, if the trend is not reversed then there would be irreversible changes leading to worst impacts of climate changes.

Natural Processes of Climate Change[5, 6]

The ***natural*** climate change includes processes, such as albedo, cloud forcing, glaciations, ocean variability, orbital variations, global forcing, radiative forcing, solar variation, and volcanization. Some of these processes are discussed briefly in the following paragraphs.

i. **Albedo:** This is the ability of a surface to reflect light that falls upon it. Snow, ice and white surfaces reflect 100% of sunshine from the surface thus having cent per cent albedo. Black bodies absorb 100% light and consequently have zero Albedo. Ice sheets reflect light fully and thus grow bigger with increasing snow fall. As an insulator, ice keeps the earth beneath it warmer. This causes the insulator to lose its grip on the land, and slide into the sea thereby continuing the process.

ii. **Cloud Forcing:** The term "cloud radiative forcing" refers to the effects clouds on both sunlight and heat in the atmosphere. Cloud radiative forcing measures how much clouds modify the net radiation of the earth, at wavelengths ranging from 0.3 to 100 micrometers. Cloud forcing (sometimes described as *cloud radiative forcing*) is the difference between the radiation budget components for average cloud conditions and cloud-free conditions. Much of the interest in cloud forcing relates to its role as a *feedback* process in the present period of global warming. All *global warming models,* used for climate change projections, include the effects of *water vapor* and *cloud forcing*. Water vapor alone contributes between 36-70% of the greenhouse effect, and when combined with clouds the total contribution increases to 66-85%.

Trapping of the long-wave radiation due to the presence of clouds, reduces the radiative forcing of the greenhouse gases, when compared to the clear-sky forcing. However, the magnitude of the effect due to clouds, varies for different greenhouse gases. Relative to clear skies, clouds reduce the global mean radiative forcing due to GHGs, such as, (i) CO_2 by 15%, (ii) CH_4 & N_2O by 20%, and (iii) halocarbons by 30%. Clouds play a significant role in our world's energy balance — they exert both a cooling effect on the surface by reflecting sunlight back into space, and a warming effect by trapping heat emitted from the surface

The magnitude of cloud radiative forcing (in Watts/m ^{2}) for a given month can be positive or negative. The positive and negative effects in the magnitude of cloud radiative forcing varies between (-) 280 to (+) 280 W/m^2 in different regions. Regions of positive cloud radiative forcing indicate areas where clouds act to increase net energy into the Earth system (*i.e.*, regions of deep tropical convection). Whereas areas of negative cloud radiative forcing, signifies regions where clouds act to decrease net energy into the Earth system (such as areas of stratus clouds off the coast of California).

Clouds remain one of the largest uncertainties in future projections of climate change by global climate models. This is due to the physical complexity of cloud processes, and the small scale of individual clouds relative to the size of the model computational grid.

iii. **Glaciation:** *Glacier* is a large, slow moving river of ice, formed from compacted layers of snow, that slowly deforms and flows by gravity. The process of glacier formation and growth is called *glaciation.* Glacier ice is the largest reservoir of "fresh water" on Earth, and second only to oceans as the reservoir of total water. Glaciers cover large areas of Polar Regions, and also form on the highest mountains in the tropics. A large quantity of *precipitation* gets trapped in glaciers, instead of flowing immediately back to the oceans. This causes the sea level to drop, and greatly modifies the hydrology of streams, and the Earth's hydrologic cycle (see chapter 14). Prolonged glaciations may cause a reduction in shallow seas, which is the most productive regions of marine products, leading to strains in marine environment.

iv. **Ocean Variability:** By using ocean simulations, from a twenty-year time series, scientists are trying to understand the causes of the variability in global ocean circulation patterns. They are also trying to determine if this variability can be predicted. The ocean's circulation can be predicted through the use of satellite-measured SSH (sea surface height), and it has been found that sea level at one location influences the variability downstream. Ocean variability leads to natural disasters like El Niño.

v. **Orbital Variation:** The Earth spins around an axis that is not perpendicular to the plane in which the Earth orbits the Sun. This tilt results in seasons. At the height of the Northern Hemisphere winter, the North Pole is tilted away from the Sun. While in the summer, it is tilted toward the Sun. The angle of the tilt varies between 22° and 24.5° on a cycle of

41,000 years. When the tilt angle is high, the Polar Regions receive less solar radiation than normal in winter and more in summer. As the Earth spins around its axis and orbits around the Sun, several quasi-periodic variations occur. Although the curves have a large number of sinusoidal components, a few components are dominant. Milankovitch studied changes in the eccentricity, obliquity, and precession of Earth's movements. Such changes in movement and orientation alter the amount and location of solar radiation reaching the Earth. This is known as *solar forcing* (an example of radiative forcing). Changes near the north polar area are considered important due to the large amount of land, which reacts to such changes more quickly than the oceans do.

vi. **Radiative Forcing:** RF is the relative effectiveness of greenhouse gases to restrict long-wave radiation from escaping back into space. For a particular greenhouse gas, radiative forcing is measured as the change in average net radiation (in watts per square meter) at the top of the troposphere, and depends on the wavelength at which the gas absorbs the radiation, the strength of absorption per molecule, and the concentration of the gas.

Greenhouse gases have a positive radiative forcing because they absorb and emit heat. Radiative forcing is a measure of greenhouse effect (global warming) and it is proportional to the concentration of these gases in the atmosphere. Radiative forcing energy (W/m^2) for a greenhouse gas, such as CO_2, in terms of relative radiative transfer in atmosphere, *i.e.* ÄF as a function of changing concentration, can be calculated from the simplified expression:

$$\Delta F = 5.35 \times \text{In}\frac{C}{C_0}$$

Where $C=CO_2$ concentration in parts per million by volume and C_0 = reference concentration. The above equation shows logarithmic relation between CO_2 concentration and radiative forcing. Therefore the increased concentrations have a progressively smaller warming effect. Formulas for other greenhouse gases such as methane, nitrous oxide, or CFCs are given in the IPCC reports Radiative forcing can be used to estimate a subsequent equilibrium surface temperature change, arising from that forcing with the following equation:

$$\delta Ts = \lambda RF$$

where λ = climate sensitivity in K/(W/m^2). A typical value of λ = 0.8, which gives a warming of 3K for doubling of CO_2.

In the pre-industrialization periods, the radiative forcing was almost constant due to negligible variation in the concentrations of the green house gases in the atmosphere.

RF is an important tool in determining the effects of greenhouse gases, aerosols, and clouds on climate change. It is necessary, however, to state that RF does not depict the climate response in its entirety. There are several parameters related to climate change that exist, but they are greatly variable and complex. Since RF is easy to calculate, it provides a general estimate of how the climate will respond to changes in greenhouse gas concentrations and various other agents.

vii. **Volcanization:** Normally volcanization leads to excess carbon dioxide in the atmosphere. However, some of the volcanic gases released from the Siberian trap flood basalts could have the opposite effect to the CO_2, cooling the climate instead of heating it.

Anthropogenic Heat

Anthropogenic heat is the heat generated by humans and human activity. It is defined as the heat and/or greenhouse gases released to the atmosphere as a result of human activities, often involving combustion of fuels. Sources include industrial plants, space heating and cooling, human metabolism, and vehicle exhausts.

Anthropogenic heat becomes more significant in dense urban areas and has small influence on rural temperatures. It is the main contributor to urban heat islands (6).

Other human-caused effects (such as changes to albedo, or loss of evaporative cooling) that might contribute to urban heat islands are not considered to be anthropogenic heat by this definition. Anthropogenic heat is a much smaller contributor to global warming than are greenhouse gases. Anthropogenic waste heat flux can be significantly high in certain urban areas. For example, waste heat flux was +0.39 and +0.68 W/m^2 for the continental United States and western Europe, respectively.

Anthropogenic heat accounts for only 1% of the energy flux created by anthropogenic greenhouse gases. Anthropogenic climate change is therefore due to excessive greenhouse gas emissions.

Anthropogenic GHGs and Radiative Forcing

The net radiative forcing due to anthrpogenic components (fig.3.3) amounts to approximately (+)1.82W/mm.sq.(5)

The radiative forcing contribution (since 1750) from increasing concentrations of well-mixed greenhouse gases (including CO_2, CH_4, N_2O, CFCs, HCFCs, and fluorinated gases) is estimated to be +2.64 Watts per square meter, wherein more than half of the amount is due to increases in CO_2 (+1.66 Watts per square meter). Thus CO_2 makes major contribution to warming relative to other climate components. According to NOAA's (National Oceanic and Atmospheric Administration, USA) Annual Greenhouse Gas Index (AGGI), which tracks changes in radiative forcing from greenhouse gases over time, shows that radiative forcing from greenhouse gases has increased 21.5% since 1990 as of 2006. Much of the increase (63%) has resulted from the contribution of CO_2. Methane is more effective greenhouse gas than carbon dioxide, but its radiative forcing is 1/4th of carbon dioxide due to presence of lower percentage.

The accumulation of excessive greenhouse gases in the atmosphere has resulted from:

i. **Fossil fuel burning:** Most of the energy requirements are met by burning fossil fuels. Fossil fuel burning has produced about three-quarters of the increase in CO_2 from human activity over the past 20 years.

ii. **Destruction of Biomes:** Biomes, like forest and ocean, act like a buffer to maintain the global temperature at a nearly constant level. However, deforestation, land-use change, destruction of marine life, lead to imbalances in eco cycles, such as,carbon-cycle and hydrological cycle. The reduction in the capacity to act as a buffer, leads to excess carbon dioxide in the atmosphere, thereby causing more heat absorption and consequent warming. An increasing number of groups worked up models that coupled climate changes with changes in soils, vegetation and the oceans. It was a matter of simple physics that as the oceans grew warmer; the sea water would absorb gases less readily. Tropical forests would show a similar effect for biological reasons. The worst worry was the Amazon forest, sustained by rains that were largely water evaporated from the jungle itself. It appeared that warming would make it harder for the planet to take up carbon, and might even trigger increased emissions. If the tropical forests get dried they would start to emit massive amounts of CO_2. would turn from a net absorber to a major emitter of carbon, speeding up climate change

Potential Effects of Anthropogenic Climate Change

The anthropogenic climate change can have disastrous consequences, and can cause frequent occurrences of severe weather. Global warming shall make a vast change in the natural climate pattern with increasing frequencies of disasters.

Summary

i. Charles David Keeling was the first to firmly establish the fact that global warming is caused by an increase in carbon-dioxide in the atmosphere. He achieved this by collecting consistent data for several decades.

ii. Keeling measured carbon dioxide content in the atmosphere and presented his data in the famous 'Keeling curve' of carbon dioxide content vs. years. His measurements at Mauna Loa, extended to 2006, shows a steady increase in average CO_2 in atmosphere from about 315 ppmv in 1958 to over 380 ppmv by the year 2006. This is a 65 ppmv in last 48 years

iii. The increase in carbon dioxide in the atmosphere is primarily due to the burning of fossil fuels for our energy needs. Global temperature rise is attributed to the absorption of higher quantities of infra-red radiation by the increasing amount of carbon dioxide in the atmosphere.

iv. While the atmosphere absorbs the infra-red portion, the rest is reflected back into space. Balance between these two determines the average temperature. Radiation balance can be altered by intensity of solar energy, variations in the Earth's orbit around the sun, volcanic eruptions etc. These are known as external forcing. These climate variation factors cannot be controlled by us.

v. In addition to natural causes of climate variation, the global warming due to anthropogenic or man-made gases in the atmosphere, can lead to drastic changes in climates with disastrous effects.

REFERENCES

1. Michael Tennsen, *Global Warming*, Alpha Books, NY, USA, 2004
2. Roger Revelle at the Scripps Institution of Oceanography, Revelle, Roger (1966). "The Role of the Oceans." *Saturday Review*, 7 May, p. 41?, http://www.aip.org/history/climate/co2.htm
3. Robert Kunzig & Wallace Broecker, *Fixing Climate –The story of climate science –and how to stop global warming*, Profile Books Limited, London,2008
4. Scripps Institution of Oceanography, CA, USA, Home of Keeling curve, http://scrippsco2.ucsd.edu/ home/index.php

5. Hegerl, Gabriele C.; *et al.* (2007). "Understanding and Attributing Climate Change" (PDF). *Climate Change 2007: The Physical Science Basis. Contribution of Working Group I to the Fourth Assessment Report of the Intergovernmental Panel on Climate Change*. IPCC. http:// ipcc-wg1.ucar.edu /wg1/ Report/ AR4WG1_Print_Ch09.pdf
6. IPCC, 2007: Summary for Policymakers; In 'Climate Change 2007; The Physical Source Basis; Contribution of the Working Group 1 to the 4th Assessment Report of the IPCC,Cambridge University Press, Cambridge, UK, http://www.ipcc.ch/pdf/assessment-report/ar4/wg1/ ar4-wg1-spm.pdf
7. Mandira Chatterjee: Urban Heat Island and Climate Change; National Seminar on Dynamics of Cities and City-Regions, Development and Environmental Issues in Contemporary India, 20-22 January, 2010,organised by Dept of Geography, Mumbai University, & The Bombay Geographical Association

6

Greenhouse Gases and Climate Change

The Earth is surrounded by a layer of gases, known as the atmosphere. Hot solar beams travel through atmosphere and reach the Earth. These beams make the Earth warm, and heat from the earth travels back into the atmosphere. There are some gases in the atmosphere which trap the heat escaping from the Earth and stop it from traveling back into space.

The discovery of atmospheric gases, with high heat absorption capacities, long life, and their increasing presence due to industrialization resulting in global temperature increase has paved the way for the current hype on global warming, and its disastrous effect on climate.

Discovery of Greenhouse Gases[1,2]

Nearly two centuries ago, in 1827, Jean Baptiste Joseph Fourier, a famous French mathematician was the first to find that some of the gases in our atmosphere behave in a manner similar to a glass-house or green-house, by allowing sunlight but preventing the solar heat radiating back in the outer space. These heat absorbing gases which cause a rise in atmospheric temperature are known as *greenhouse gases*.

John Tyndall, a British physicist in 1850's, found that the heat absorption capacity of carbon dioxide, methane, water vapor and few other gases is enormous. They constitute less than 1% of the atmosphere, but their superior heat absorption capacity keeps the earth warm. Nitrogen and oxygen, which form 99% of the dry air, do not have the capacity to absorb heat.

Greenhouse gases make the Earth warm, and without that it would be too cold for living things, such as plants and animals. Naturally occurring greenhouse gases follow the natural cycles of absorption and emission.For example, carbon dioxide, a major greenhouse gas is formed and absorbed by natural 'carbon cycle'. Carbon cycle maintains the same level of carbon dioxide in the atmosphere. However, man-made greenhouse gases are being produced everyday by transport systems, power generation and other manufacturing industries. These industrial emissions due to human activities are known as anthropogenic greenhouse gases. Large amount of anthropogenic gas emissions, makes the carbon cycle unbalanced and results in higher residual content of greenhouse gases in the atmosphere.The larger is the quantity of greenhouse gas in the atmosphere, the higher is the temperature rise.

Greenhouse gases, their origins and sinks, life cycles, global warming potentials and other characteristic properties are discussed in this chapter.

Infra-red Radiation and the Greenhouse Effect[3, 4]

The amount of energy emitted by radiation at temperature T is given by the Stefen-Boltzmann equation (see chapter3) modified for the effects of emissivity, e, as follows:

$$E = ex. T^4$$

Where σ = Stefan – Boltzmann constant $=5.67 \times 10^{-8}$ W m^{-2} K^{-4}

The variation of emissivity with wavelength is small for solid objects and is almost equal to1.0 or more than 0.90 for various types of the Earth's surface. However, he emissivity of gases varies over a wide range. Each gas absorbs radiant energy over a narrow wavelength range, called *spectral absorption lines*. Normally these lines are grouped together as *absorption bands*. The location and the absorbing capacity depend on the molecular structure of the gas. With the increase in the amount of absorbing gas, its temperature, and the total atmospheric pressure, these bands are broadened and the amount of absorption increases.

The atmosphere is highly absorbent to infra-red radiation band due to the effects of water vapor, carbon dioxide and other trace gases. The opaqueness of the atmosphere to infra-red radiations, relative to its transparency to short wave visible spectrum, is commonly referred to as the '*greenhouse effect*' The gases showing opaque behavior to infra-red radiation through absorption are called *greenhouse gases.*

Greenhouse Gases, Natural and Anthropogenic

Some of the greenhouse gases like carbon dioxide (CO_2), methane (CH_4), nitrous oxide (N_2O), belong to both natural and anthropogenic origins, while others like moisture (H_2O) & ozone (O_3), are of natural origin.The *atmospheric water vapor (H_2O) also makes a large contribution to the natural greenhouse effect but it is thought that its presence is not directly affected by human activity.* Unlike other GHGs, moistures do not have the anthropogenic origin. Hence there is no enhanced greenhouse effect due to non-existent anthropogenic moisture in the atmosphere. Other gases like, sulfur dioxide (SO_2) chlorofluorocarbon (CFCs) do not occur in the nature, but are of only anthropogenic(man-made) in origin.

Shortwave solar radiation can pass through the clear atmosphere relatively unimpeded. Longwave infra-red radiation, on the other hand, is emitted by the warm surface of the Earth and is absorbed partially and then re-emitted by a number of trace gases (particularly water vapor and carbon dioxide, which are present in the cooler atmosphere above). It is necessary to distinguish between the "natural" and a possible "enhanced" (formed due to anthropogenic) greenhouse effect. The natural greenhouse effect *i.e.*, the greenhouse gases produced by natural cycles, can cause the mean temperature of the Earth's surface to rise up to about 33° C. Natural greenhouse effect creates a climate in which life can thrive and sustain. Without the heating effect of natural greenhouse gases, the Earth would be a very frigid and inhospitable place. However, an enhanced greenhouse effect refers to the possible increase in the mean temperature of the Earth's surface above and beyond that occurring due to the natural greenhouse effect. The large amount of greenhouse gases in the atmosphere beyond that due to natural process, belong to the anthropogenic origin. Increase in global warming would probably bring other, sometimes deleterious changes in climate, such as, changes in precipitation, storm patterns, and the level of the oceans.

According to the Intergovernmental panel on Climate Change (IPCC), a joint project of the World Meteorological Organization and the United Nations Environment Program(5), if the carbon dioxide in the atmosphere continues to rise, the world will warm up by 2.5° F to 10.4°F (1.4 to 5.8°C) by the end of 21st century. With a global temperature rise of 1.4°C, there would not be a major problem due to climatic changes. However if the temperature goes up to the highest level of 5.8°C, there would be a dramatic change in climate with disastrous consequences across the world.

The greenhouse effects of major greenhouse gases(GHGs) include following:

i. Water Vapor → causes ~ 36–70% of the GHG effect (not including cloud)

ii. Carbon Dioxide (CO_2), → causes~ 9–26% of GHG effect

iii. Methane (CH_4) → causes~ 4–9% of GHG effect

iv. Ozone(O3)→ causes ~3–7%.of GHG effect

Other gases contribute very small fractions of the greenhouse effect. However, in the recent decades there has been an increasing concentration of Nitrous Oxide (N_2O) as a result of human activities, such as, the use of nitrate fertilizer in agriculture.

The atmospheric concentration of CO_2 and CH_4 has increased by 36% and 148% respectively since the beginning of the industrial revolution in the mid-1700s. The present CO_2 levels are considerably higher, than at any time during the last 650,000 years, the period for which reliable data has been extracted from ice core. Also almost all the increase is due to human activities(5).

Properties of Greenhouse gases

The major properties of greenhouse gases, their origins and sinks, and the effects on global warming are tabulated in Table 5.1.

1. Radiative Forcing[5]

This is a measure of greenhouse effect (global warming) of the greenhouse gas, and is proportional to its concentration in the atmosphere. Molecule for molecule, methane is a more effective greenhouse gas than carbon dioxide, but its concentration is much smaller so that its total radiative forcing is only about a fourth of that from carbon dioxide (see chapter 3 & 4).The estimated average radiative forcing figures for greenhouse gases in 2005 are as follows:

CO_2 = 1.7W/sqm,

CH_4 = 0.5 W/sq.m

N_2O = 0.15 W/ sq. m

Halocarbons -0.4 W/sq.m

2. Global Warming Potential (GWP)[6,7]

The Global Warming Potential (GWP) is used within the Kyoto Protocol as a measure for the climatic impact of emissions of different greenhouse gases. A GWP value is defined over a specific time interval, so the length of this time interval must be stated to make the value meaningful. For example, methane has a GWP of 72 over 20 years, but a lower GWP of 25 over 100 years. This is because it is very potent in the short-term but then breaks down to CO_2 and water in the atmosphere in the longer the period.

GWP is based on a number of factors, including the *radiative efficiency* (heat-absorbing ability) of each gas relative to that of carbon dioxide, as well as the *decay rate* of each gas (the amount removed from the atmosphere over a given number of years) relative to that of carbon dioxide. It is a relative scale which compares the gas in question to that of the same mass of carbon dioxide, whose GWP is by definition equals to 1. A GWP is calculated over a specific time interval and the value of this must be stated whenever a GWP is quoted or else the value is meaningless. GWP is defined by the IPCC[6] as the ratio of the time-integrated radiative forcing from the instantaneous release of 1 kg of a trace substance relative to that of 1 kg of a reference gas:

$$GWP(x) = \frac{\int_0^{TH} ax.[x(t)]dt}{\int_0^{TH} ax.[r(t)]dt}$$

where:

The term TH is the time period which over value is integrated. The choice of the time horizon depends on whether to emphasize shorter-term processes (*e.g.*responses of cloud cover to surface temperature changes) or longer-term phenomena (such as sea level rise) that are linked to sustained alterations of the thermal budget (*e.g.*, the slow transfer of heat between, for example, the atmosphere and ocean). In addition, if the speed of potential climate change is of greater interest rather than the eventual magnitude, then a focus on shorter time horizons can be useful[7].

The term a_x is the *radiative efficiency* due to a unit increase in atmospheric abundance of the gas (*i.e.*, Wm^{-2} kg^{-1}) and [x(t)] is the time-dependent decay in abundance of the substance following an instantaneous release of it at time. The radiative efficiencies a_x and a_r are not necessarily constant over time. The time horizon plays a role in the GWP values. The parties to the UNFCCC have also agreed to use GWPs based upon a 100 year time horizon. The GWP is actually calculated in terms of the 100 year warming potential of a kilogram (kg) of a gas relative to that of a kilogram of CO_2.(tab.3.1). According to the IPCC, GWPs typically have an uncertainty of ±35 percent

The absorption of infra-red radiation by many greenhouse gases varies linearly with their abundance. A few important ones display non-linear behavior for current and likely future abundances (*e.g.*, CO_2, CH_4, and N_2O). For those gases, the relative radiative forcing will depend upon abundance and hence upon the future scenario adopted. Since all GWP calculations are a comparison to CO_2, which is non-linear, all GWP values are affected.

Under the Kyoto Protocol, the Conference of Parties decided (decision 2/CP.3) that the values of GWP calculated for the IPCC Second Assessment Report are to be used for converting the various greenhouse gas emissions into comparable *CO_2 equivalent* when computing overall sources and sinks.

The GWP has been subjected to many criticisms because of its formulation, but it has retained some favor because of the simplicity of its design and application, and its transparency compared to proposed alternatives.

3. Carbon Dioxide Equivalent

Carbon dioxide is the major anthropogenic and natural greenhouse gas, which plays a key role in global warming. The effect of other gases on global warming are thus evaluated in comparison to carbon dioxide as GWP.GWP itself is a measure for comparing the global warming potential of various greenhouse gases in comparison to carbon dioxide. Carbon dioxide equivalents are commonly expressed as "*million metric tons of carbon dioxide equivalents (MMTCO2Eq).*" The carbon dioxide equivalent for a gas is derived by multiplying the million metric tons of the gas by the associated GWP (8), as follows:

MMTCO2Eq = (million metric tons of a gas) × (GWP of the gas)

Carbon dioxide equivalents provide a universal standard of measurement against which the impacts of releasing (or avoiding the release of) different greenhouse gases can be evaluated.

As there is no possibility of directly influencing atmospheric water vapor concentration by anthropogenic sources, hence the GWP-level for water vapor is not calculated.

4. Global Temperature Change Potential (GTP): New Metrics for GWP[9]

Two new metrics, based on a simple analytical climate models have been proposed for global warming potential (GWP). These new metrics are: i) *Global Temperature Change Potential,* which is the temperature change at a given time due to a *pulse emission of a gas (*GTP_P); ii) *Global Temperature Change Potential due to Sustained Emission Change,* which is the temperature change due to a *sustained emission change* (hence GTP_S). Global Temperature Change Potential (GTP) goes further than GWP and integrated RF in describing the effects of emissions.It estimates the change in global mean temperature for a selected year in the future. Both GTP_P and GTP_S are calculated relative to the temperature change due to a similar emission change of a reference gas, which is ubiquitous carbon dioxide.

GTP is based on RF. However GTP calculation is more complex because it calculates climate response and not just radiative forcing. GTP accounts for Earth's thermal inertia, *i.e.*, the lag between when the emissions occur and when they cause warming. The calculation takes into account also the Earth's climate sensitivity rather than simple radiative forcing calculations. It is more useful for policy makers to know what the actual temperature change will be than only the amount of energy that has been added to the system.

When compared against an *upwelling-diffusion energy balance model* that resolves land and ocean and the hemispheres, GTP_P shows poor performance (except for long-lived gases) but GTP_S shows good performance, for gases with a wide variety of lifetimes. Also *for time horizons in excess of about 100 years, the* GTP_S *and GWP produce very similar results, indicating an alternative interpretation for the GWP.*

GTP can be used to express future climate responses to current aviation emissions. As with GWP, the chosen time horizon greatly influences the results. Short time horizons include the warming due to short-lived emissions, whereas longer time horizons exclude those effects.

5. Life Span of GHGs in Atmosphere

The longer a greenhouse gas stays in the atmosphere, the more its cumulative heating effect. Although the GWPs are calculated on a 100 years basis, lifetime becomes an important parameter for the longer term heating effect.

Carbon dioxide has a long lifetime 100 years. However it remains in the atmosphere for several thousand years, but the impact on global warming is reduced to less than one quarter of its original impact after the first 100 years. The life span of nitrous oxide is 114 years. In contrast, both methane and HFCs are quickly removed from the atmosphere due to their relatively short atmospheric lifetimes of around 10 years. As a major emission, the long lifetime of CO_2 means a significant commitment for climate change, long into the future.

6. Climate Impact

The total quantity released is as important as the GWP in calculating the real environmental impact.

IMPACT= INDEX × QUANTITY

Despite the low GWP of CO_2, the enormous quantities emitted and its long lifetime means that it has a far greater impact on climate than any other greenhouse gas. Currently, *CO_2 emissions contribute 64% of the total greenhouse gas emissions and the figures are still rising.*

7. Feedbacks

One of the most pronounced feedback effects relates to the evaporation of water. Warming by greenhouse gases, such as, CO_2 will cause more water to vaporize into the atmosphere. Since water vapor itself acts as a greenhouse gas, the atmosphere warms further; causing more water vapor formation, and the process continues until a new dynamic equilibrium between water/ vapor is reached. This results in a *larger greenhouse effect than that due to CO_2 alone.* Although this feedback process causes an increase in the absolute moisture content of the air, the relative humidity stays nearly constant or even decreases slightly because the air is warmer. This feedback effect can only be reversed slowly as CO_2 has a long average atmospheric lifetime.

Individual Greenhouse Gases

The annex A of the Kyoto Protocol, a UN convention on climate change, includes the following greenhouse gases in the list:

Carbon dioxide (CO_2)

Methane (CH_4)

Nitrous oxide (N_2O)

Hydro fluorocarbons (HFCs)

Per fluorocarbons (PFCs)

Sulfur hexafluoride (SF_6)

Ozone is covered by Montreal Protocol, hence omitted. Kyoto Protocol makes it obligatory for the signatory countries to control emissions of these gases to certain permissible limit. The properties of greenhouse gases are given in tab.5.1[5, 6, 7, 10].

Table 5.1: Greenhouse Gases, Origin, Sink, Life, and GWP*

GHGs	*Origin*	*Sink*	*Impact on Climate*	*Life in years*	*GWP*** 100yrs*
i.Carbon dioxide CO_2	I burning fossil fuel ii. Deforestation	i. ocean ii. Plants (photo synthesis)	absorbs infrared (heat) radiations makes atmosphere warem too much cause global warming key to plant life affects stratosphere O absorbs infrared, affects stratosphere & troposphere O & OH. Produces Co	80-100	
ii. Methane CH_4	i. bio-mass burning ii. swamps, wetlands ii rice paddies, iii rice paddies, iv. landfill, caracas backterial attack	i. reaction with OH ii. Micro-organisms in soil		10-12	23
iii.Nitrous oxide, N_2O	i.biomass burning ii. Fossil fuel burning iii. nitrate fertilizer	i.soil (photolysis)& reaction with O	Absorbs infrared, affects stratosphere O	114	296

iv. Hyderoflu- -rocarbon HFCs	i.refrigeratio leakage	—	absorbs infrared	16-41	1200
v. Perfluoro- carbon PFCs	i.semi-conductor &electronic industries. Leakage	—	absorbs infrared	3200- 50000	6500 11,200 (100yrs)
vi. Sulfur hexafluo- -ride, SF6	i.switch=gears		absorbs infrared	3200	24,900
vii. Sulfur dioxide SO_2	i.volcanos ii. Burning coal, kerosene	reacts with H2O forms weak acid	forms aerosols, affects ozone layer		

*Based on various sources, including IPCC reports & GWP on intergovernmental Panel on Climate Change, 3rd agrrement report, 2011

** compared to CO_2=1

*i. **Carbon Dioxide***

The original atmosphere was that due to carbon dioxide from the volcanic emissions. It is only through the evolution of plant life, which began to take in carbon dioxide and give up oxygen that the earth's atmosphere changed to one containing mostly oxygen, with a small share of carbon dioxide. However, burning of fossil fuels, led to an increase in carbon dioxide from 280 ppm, a stable figure before industrialization, to 367 ppm in recent times. The net effect is a high rate of rise in the atmospheric carbon dioxide, which if not reduced, can lead to carbon dioxide concentration figures in the range of 490 to 1200 ppm in 2100, as predicted by Scientists at IPCC.

An important function of carbon dioxide is to sustain plant life through formation of carbohydrates by photosynthesis. This is a biochemical process using light energy to make sugar from CO_2, and H_2O, while releasing O_2 as per the following chemical reaction:

$$6CO_2 + 6H_2O \text{ (+ light energy)} = C_6H_{12}O_6 + 6O_2.$$

This is the source of the O_2 we breathe, and thus, a significant factor in the concern about deforestation. This process occurs in plants and some algae (photo planktons).This reaction doesn't directly need light in order to occur, but it does need the products of the light reaction (ATP and another chemical called NADPH). The dark reaction takes plăce in the storma involves a cycle called the ***Calvin cycle*** in which CO_2 and energy from ATP are used to form sugar. Stroma is a chlorophyll-containing plastid found in algal and green plant cells.

Another sink for CO_2 is the ocean. Scientists felt that the ocean being a sink for carbon dioxide, would absorb the extra carbon dioxide. However, Scripps Institute of Oceanography found that the *carbon dioxide absorption capacity of sea water is limited.*[11]

A unique feature of carbon dioxide gas is the long life span of about 100 years in the atmosphere. The cascading effect of increasing carbon dioxide through accumulation over 100 years would lead to an alarming rise in global atmospheric temperature. The long life of carbon dioxide, also means that even if we don't burn fossil fuel for next 100 years, the present level of carbon dioxide would remain at the same level in 2100.

According to IPCC, in order to stabilize the carbon dioxide level in the atmosphere at 450 to 1000ppm, global use of coal, oil, and gas will have to be reduced to below the level at 1990, and continue to decrease steadily thereafter to a small fraction of whatever they are today.

Other Greenhouse Gases

Although their levels in the atmosphere are much lower than that of CO_2, gases like methane and fluorinated gases are also potent greenhouse gases.

ii. Methane

Methane not only absorbs infra-red radiation, but also affects troposphere & stratospheric ozone. Methane on burning produces carbon dioxide. Although methane has higher global warming potential compared to carbon dioxide, due to its lower life ($1/10^{th}$ of CO_2) and concentration its global warming effect is much less than that of carbon dioxide. Sinks for methane includes reaction with OH, and microorganisms in soil. Although methane ("marsh gas") is released by natural processes (*e.g.* from decay occurring in swamps), human activities may now account for over one-half of the total. The sources include, growing rice in paddies, burning forests and in raising cattle, In raising cattles, the fermentation in their rumens produces methane that is expelled — collectively adding an estimated 100 million tons a year to the atmosphere. It was reported that some plants naturally release methane to the atmosphere. On re-examination, it was found that methane emissions from plants are negligible and do not contribute to global climate change[12].

The burning of the tropical rain forest adds to the atmospheric methane budget. The methane concentration in the atmosphere is presently some 1.8 ppm and is growing at a rate of 1% per year. Although this concentration is far less than that of CO_2, methane is 30 times as potent a greenhouse gas and so may now be responsible for 15–20% of the predicted global warming.

iii. Nitrous Oxide(N_2O)

Nitrous oxide is produced by burning biomass, fossil fuels, and from nitrate fertilizers. Nitrous oxide absorbs infrared radiation and affects stratospheric ozone. Nitrous oxide has a very long life of 114 years. Also its global warming potential is very high, in the regions of 275 (for 20years), 296(100years) and 197 (in 500 years) when compared to that of CO_2, which has a GWP of 1 (tab.3.1).

iv. High GWP Gases- Fluorinated Gases[6,13]

There are three major groups or types of high GWP fluorinated gases: hydrofluorocarbons (HFCs), perfluorocarbons (PFCs), and sulfur hexafluoride (SF_6).All of them are of anthropogenic origin.

a. **Hydrofluorocarbon(HFCs):** HFC's stand out as a good choice for refrigeration due to their ODP(Ozone Depletion Potential) of 0, but have a high GWP of 1200 and an atmospheric lifetime of 16 years. Pentafluoroethane (HFC-125) degrades in the atmosphere to CO_2 and HF by reaction with naturally occurring hydroxyl radicals. The atmospheric lifetime is estimated to be 40.7 years. Pentafluoroethane, has no effect on stratospheric ozone since it contains neither chlorine nor bromine. However, Pentafluoroethane has a global warning potential of 2800 relative to CO_2 at 100 years.

b. **Perfluorocarbons (PFCs):** A PFC (perfluorinated carbon) compound is defined as a compound containing carbon and fluorine only. Perfluorocarbons are highly volatile, linear, branched chain or cyclic per-fluorinated carbons (C1 up to C6, fully saturated).This definition covers the most common commercial PFC gases and liquids with boiling points up to 56°C even though there are other PFC compounds, mainly used in closed system heat transfer applications.

The GWP and Lifetime data of some PFCs are as follows:

Perfluorocarbons	GWP	Lifetime
Perfluoromethane	6,500	50,000
Perfluoroethane	9,200	10,000
Perfluorocyclobutane	11,200	3,200
Perfluoropentane	8,900	4,100

PFC gases and liquids are traditionally used in several electronic industry processes ranging from semiconductor front-end manufacturing, IC-component quality control testing, to direct contact dielectric cooling of power electronics assembly. With the phase-out of ozone-depleting substances, PFCs have been introduced. However, PFCs are only selected for"high-end" performances, where system efficiency and worker safety are mandatory due to extremely high price (>60 Euro/l).

c. Sulfur Hexafluoride (SF6)

Sulfur hexafluoride (SF_6) is an excellent dielectric gas, which is used for high and medium voltage switchgear. It is chemically inert, gaseous even at low temperature, non flammable, non toxic, non corrosive. Its combined chemical, thermal and electrical properties allow many advantages to be achieved. The global warming properties of SF6 include the following:

*GWP = 24,900(100 years); *Lifetime = 3,200 years

SF6 is present in air in its inert form.

Ecological limits dictate that *to avoid dangerous climate change we must reduce total greenhouse gas emissions below 1990 levels by a minimum of 50% within the next fifty years, and 60-80% in the next century.*

Other Minor Gases

Carbon Monoxide(CO): This is an unstable gas, emitted by industries and transports, while burning carbon containing fuels. It reacts with oxygen in air to produce CO2,and is also absorbed by soil.

Sulfur Dioxide: It originates from volcanic eruptions and is also produced by burning coal and biomass. The sinks for SO_2 include dry and wet depositions and reaction with moisture. It forms aerosols, which scatter solar radiation.

Ozone Layer Depleting Gases

The Ozone layer in the stratosphere filters the harmful UV-rays from solar radiation before it reaches to Earth's atmosphere. However there are several industrial gases, which are found to be responsible for depleting ozone in the ozone layer. The major ozone depleting gases are CFC, HCFC and HFC. Ozone depletion capacity of a gas is expressed by the Ozone Depletion Potential (ODP), and depends like the GWP on the atmospheric lifetime of the gas (measured in years). Atmospheric lifetime is a critical factor, as it determines how long the substance in question can continue to cause damage.

Table 5.2 Comparative Properties of CFCs and HFCs

Gas		*Lifetime (years)*	*ODP*	*GWP (years)*	
				20	*100*
Chlorofluocarbon					
CFC-11	CCI3F	45	1.0	6300	4600
CFC-12	CCI2F2	100	0.82	10200	10600
CFC-13	CCIF3	640		10000	14000
Hydrofuororcbons					
HFC-23	CHF3	260	<0.004	9400	12000
HFC-32	CH2F2	5.0		1800	550
HFC-41	CH3F	2.6		330	97

a. *Chlorofluorocarbons (CFCs)*: These are synthetic gases in which the hydrogen atoms of methane are replaced by atoms of fluorine and chlorine (*e.g.*, CHF_2Cl, $CFCl_3$, CF_2Cl_2). CFCs absorb infrared and dissociate and react with ozone in stratosphere. CFCs have long life spans from 45 years for CFC_{11}, through 100 (CFC_{12}), 640 (CFC_{13}) to 1700 (CFC_{15}).Global warming potential (GWP) for CFCs have a range of 4500 to 10,200 when compared to carbon dioxide. Their long lifetime means that they can be transported slowly to the stratosphere. Although CFC's were banned within a few years of the Montreal Protocol in 1987, their long lifetime means that it will take many more years before their atmospheric concentrations level out and then decrease.

b. *HCFC:* HCFC's are an improvement over CFC's with a GWP of 1500, an ODP of 0.07 and an atmospheric lifetime of 15 years. As a refrigerant, they still fall short of environmental safety.

Steps to reduce halocarbon generation

Unlike other GHGs, the halocarbons are totally man-made products, hence completely avoidable with the development of suitable alternative chemicals.

Most HCFC's are now banned in developed countries (Montreal Protocol and EU Regulation). They are still allowed in developing countries. HFC/PFC/SF_6 emissions present a real danger to the planet. Eliminating the use of HFCs/PFCs/SF_6 is one of the easiest ways for advanced nations to accomplish their commitments under the Kyoto Protocol and for companies to demonstrate corporate environmental leadership. Environmentally safer, cost effective, technically reliable alternatives to HFCs/PFCs exist in virtually all applications. There are companies in many countries that are ready to provide these alternative technologies.

Summary

i. Greenhouse gases have the property of absorbing infra-red radiation of solar beams thus causing rise in the atmospheric temperature. The list of main GHGs includes carbon dioxide, methane, nitrous oxide, hydrofluorocarbons, perfluorocarbons, and sulfur hexafluorides.

ii. The heating potential of GHGs are evaluated in comparison with that of carbon dioxide's heating potential, and the value expressed as global heating potential, or GWP.

iii. GHGs can stay in the atmosphere for long time, may be 100 years or more. Hence GWP values are estimated for 100 years duration in the atmosphere. Also the amount of gas present has a direct impact on heating potential, hence impact on atmosphere is determined by multiplying GWP with the amount present.

iv. Some of the GHG, mainly fluorocarbons, such as CFCs can destroy ozone layer in the stratosphere. However, these are of anthropogenic origin of limited industrial uses. Their uses in industries are banned by the countries signing Montreal Protocol, and with the use of suitable substitutes the ozone depletion problem has been minimized.

REFERENCES

1. Michael Tennsen, *Global Warming*, Alpha Books, NY, USA, 2004.
2. Robert Kunzig & Wallace Broecker, *Fixing Climate –The story of climate science –and how to stop global warming*, Profile Books Limited, London, 2008.
3. R.G. Berry and R.J.Chorley, *Atmosphere, Weather, & Climate*, Chapter2, Solar Radiation and Global Energy Budget, pp. 20-50, Routledge, NY, 2002.
4. Peter J Robinson & Ann Henderson-sellers: *Contemporary Climatology*. Chapter 2, The Earth's Radiation Budget, pp. 17 -38, Longman, 2nd edition, 1999.
5. 2007 IPCC Fourth Assessment Report (AR4) by Working Group 1 (WG1) and Chapter 2 of that report (Changes in Atmospheric Constituents and in Radiative Forcing) which contains GWP information.
6. 2001 IPCC 3rd Assessment Report (TAR) *Global Warming Potentials & Direct GWP*.
7. IPCC-1994, Climate Change 1994, Special Report on Emission Scenerio.
8. Environmental Protection Agency (EPA), USA, *Glossary of Climate Change Terms*, http://www.epa.gov/climatechange/glossary.html)
9. Carbon Offset Research & Education (CORE) is an initiative of the Stockholm Environment Institute (SEI). Global Temperature Change Potential, http://www.co2offsetresearch.org/aviation/GTP.html
10. Inventory of US greenhouse gas emissions and sinks, 1990-2007- (April 2009) Executive summary, PDF, 21 pp., 906 KB http:// www.epa.gov/ climate change /emissions/downloads09/GHG2007-ES-508.pdf
11. Roger Revelle at the Scripps Institution of Oceanography, Revelle, Roger (1966). "The Role of the Oceans." *Saturday Review*, 7 May, p. 41, http:/ /www.aip.org/history/climate/co2.htm
12. Science Daily, April, 27, 2007.
13. US Environment Protection Agency (EPA), Climate Change,High HWP gases,Science,http://www.epa.gov/highgwp/scientific.html
14. A halocarbon management –Fact sheet #4, Pollution Prevention Fact Sheet, Pollution Prevention Program, The Federal Programs Division of Environment Canada, Ontario Region, http://www.p2pays.org/ref/ 19/18372.pdf.

7

Biome and Ecology

The climate zones have their own characteristic environmental conditions. In a given climatic condition, specific types of biotic (living) and non-biotic (non-living) groups can exist in harmonious relations amongst the three, to make up, what is known as *environment*. An *ecosystem* is a geographical area of a variable size where plants, animals, the landscape and the climate all interact together. Within an ecosystem, organisms occupy areas, where physical conditions, especially climate factors such as light, heat, moisture and wind are most appropriate to cater to their needs. These areas are called *habitats*[1].

The communities of plants and animals occupying major geographical areas, (continental or sub continental in scale), that retain sufficient similarities of form and function, that are recognized as ecosystems, are called biomes. *Biomes* are defined as "the world's major communities, classified according to the predominant vegetation and characterized by adaptations of organisms to that particular environment"[2]. A biome can be thought of many similar ecosystems throughout the world grouped together. An ecosystem can be as large as the Sahara Desert, or as small as a puddle or vernal (temporary) pool.

A biome is usually considered to have the attributes of *climax community* – a community that represents the most developed combination of plants and animals possible under the environmental conditions at a given time in a given area. The *biomes are dynamic entities*, and can retain the equilibrium to a considerable degree of variations in environmental conditions such as those occurring during seasonal changes. However, *prolonged environmental disruptions, such as, that caused by climate change or fire, may alter a specific ecosystem irreversibly and bring a system with different characteristics.* As a result maps of biomes only represent the period in which it has been in existence but surely not the current scenario. If the remaining biomes are to be protected and preserved, and the projected changes threatened by global warming are to be managed, it is important to know what currently exists and what might be expected to come in a particular area. Biomes are responsible for keeping the ecological balance and maintaining of optimum quantities of carbon dioxide in the atmosphere and the climate. Hence they need to be preserved.

Types of Biomes[2,3,4,5]

There is no accepted classification of biomes. However biomes are commonly divided into terrestrial, fresh water and marine groups. The four major terrestrial biomes consisting of deserts, forests, grassland and tundra can be further subdivided into around 10 more biomes (table 6.1). On land, biomes are separated primarily by latitude.

A biome contains a large area with similar flora, fauna, and microorganisms. The familiar *tropical rainforests, tundra* in the arctic regions, and the evergreen trees in the *coniferous forests,* belong to one or other of large communities containing species that are adapted to its varying conditions of water, heat, and soil. For instance, polar bears thrive in the arctic while cactus plants have a thick skin to help preserve water in the hot desert.

Table 6.1 Types of Biomes

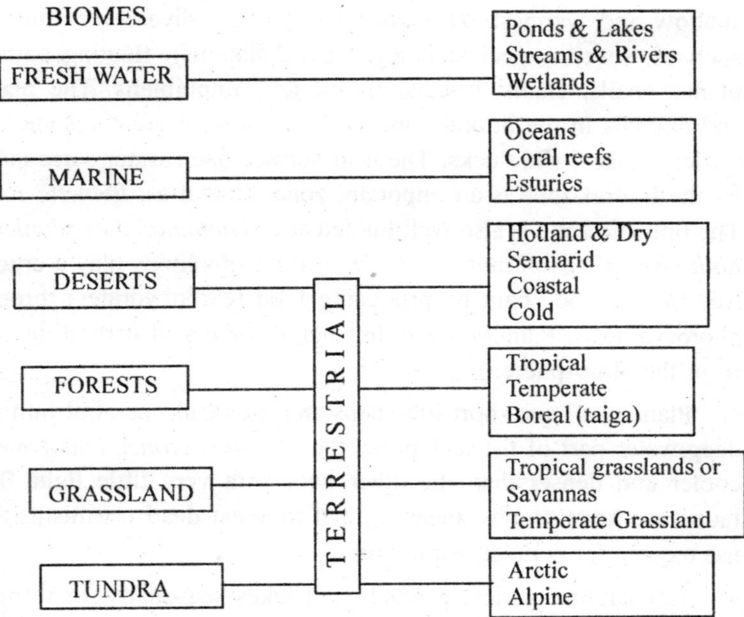

The six biomes and their characteristic features are as follows:

1. Freshwater

Freshwater is defined as water with low salt concentration (<1%). Plants and animals growing in this biome would not survive in sea water with high salt content. The three different types of freshwater regions include (i) ponds and lakes (ii) streams & rivers and (iii) wetlands.

Ponds and lakes

Small ponds to big lakes are scattered throughout the Earth. Many ponds are seasonal, while lakes may exist for hundreds of years or more. Ponds and lakes may have limited species diversity since they are often isolated from one another and from other water sources like rivers and oceans. Three different zones of ponds and lakes differ in features and species.

The topmost zone near the shore known as the *littoral zone* is shallow and warmest. This zone has a fairly diverse community, such as, several species of algae (like diatoms), floating aquatic plants, snails, clams, insects, fishes, and amphibians. The plants and animals in the littoral zone become food for creatures such as turtles, snakes, and ducks. The near-surface open water surrounded by the littoral zone is an important zone, known as, *limnetic zone*. The limnetic zone is also well-lighted and is *dominated by plankton, both phytoplankton and zooplankton.* Phytoplanktons play a crucial role in the food chain by producing food (carbohydrates) through photosynthesis. Planktons are thus net absorbers of carbon dioxide from the atmosphere.

Planktons have short life spans and the dead ones fall into the deep-water part of the lake/pond, the *profundal zone*. This zone is cooler and denser than the other two, with very little light. The fauna are heterotrophs, meaning that they eat dead organisms and use oxygen for cellular respiration.

Temperature varies in ponds and lakes seasonally. In tropical regions, the temperature at the top in summer can be as high as outside temperature of up to 35° C, with some cooling effect due to evaporation, while the bottom can remain cool at 18° C to 10° C. In between the two layers, there is *a narrow zone called the thermocline* where the temperature of the water changes rapidly.

Streams and rivers

These are bodies of flowing water moving in one direction, from their starts at headwaters, which may be springs, snowmelt or even lakes, and then travel all the way to their mouths, usually another water channel or the ocean. The cool, clear water, with high oxygen levels at the source lead to freshwater fish such as trout and heterotrophs to survive in this region. At the middle part of the stream/river, the width increases along with species diversity. Numerous aquatic green plants and algae can be found in this area. Toward the mouth of the river/stream, the water becomes murky from all the sediments that it has picked up upstream, decreasing the amount of light that can penetrate through the water. With less light, there is less diversity of flora, and lower oxygen levels allows fish that require less oxygen, such as catfish and carp, to survive.

Wetlands

Wetlands are areas of standing water that support aquatic plants. Marshes, swamps, and bogs are all considered wetlands. Plant species adapted to the very moist and humid conditions are called hydrophytes. These include pond lilies, cattails, sedges, tamarack, and black spruce. Marsh flora also includes such species as cypress and gum. *Wetlands have the highest species diversity of all ecosystems.* Many species of amphibians, reptiles, birds (such as ducks and waders), and furbearers can be found in the wetlands. Wetlands are not considered freshwater ecosystems as there are some, such as salt marshes, that have high salt concentrations, supporting different species of animals, such as shrimp, shellfish, and various grasses. Wetlands are found to be useful in

i. controlling floods,

ii. contributing to biological processes for natural resources, pollination by insects, birds, and mammals, and

iii. the most important function concerning global warming is regulating the climate by forests and open space through oxygen production and carbon dioxide consumption.

Freshwater biomes have suffered mainly from pollution. Runoff containing fertilizer and other wastes and industrial dumping enter into rivers, ponds, and lakes and tend to promote abnormally rapid algae growth. When these algae die, dead organic matter accumulates in the water. This makes the water unusable and it kills many of the organisms living in the habitat. Stricter laws have helped to slow down this thoughtless pollution.

2. Marine

Marine regions cover about three-fourths of the Earth's surface and include oceans, coral reefs, and estuaries. *Marine algae supply much of the world's oxygen supply and take in a huge amount of atmospheric carbon dioxide.* The evaporation of the seawater provides rainwater for the land. Marine regions consist of followings:

- Oceans
- Coral reefs
- Estuaries

Oceans

The *largest of all the ecosystems,* oceans are very large bodies of water that dominate the Earth's surface. The oceans regions have been divided in several separate zones (fig.6.0). All zones have great diversity of species. The ocean probably has the richest diversity of species with lesser number than on land. The characteristic features of different zones are as follows:

Intertidal zone: is where the ocean meets the land. This zone is either submerged and or exposed, by waves and tides. The submerged high tide area contains algae and small animals, such as snails, crabs, sea stars etc., while at the bottom of intertidal zone, exposed during lowest tides, fishes, & seaweed are found. In intertidal zone on sandy shores, waves keep mud and sand constantly moving, thus mostly found some fauna, such as worms, clams, crabs, and shorebirds.

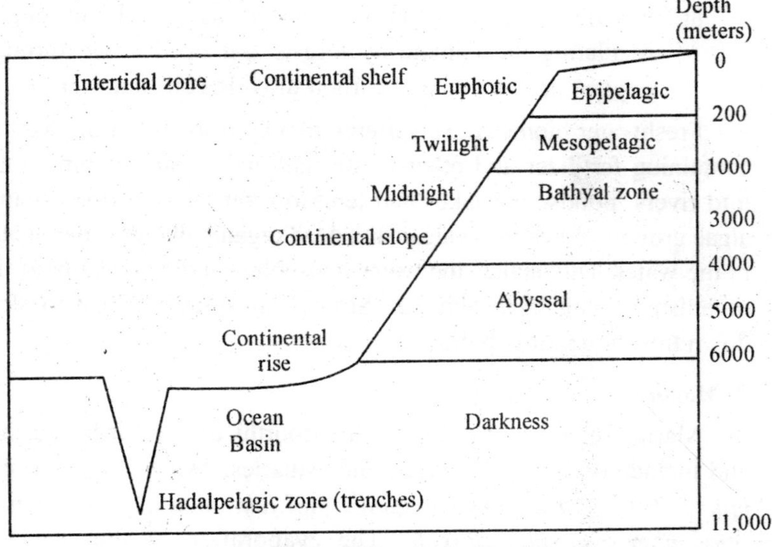

Fig. 6.0 Layers at different depth of the Ocean

Epipelagic Zone: The epipelagic zone includes those waters further from the land, basically the open ocean. This surface layer is also called the sunlight zone and extends from the surface to 660 feet (200 meters). The solar heat is responsible for the wide range of temperatures that occur in this zone, both in different latitude and each season. Surface temperature "follows the sun". From the earth's perspective, the sun's position in the sky moves higher each day from winter to summer.This change in the sun's position from winter to summer means that in summer more energy is reaching the ocean and therefore warms the water The sea surface temperature can be high at 97°F (36°C) in the Persian Gulf and low at 28°F (-2°C) near the north pole.

Interaction with the wind keeps this layer mixed and thus allows the heating from the sun to be distributed vertically. At the base of this mixing layer is the beginning of the thermocline (Fig.6.0.1). The thermocline is a region where water temperature decreases rapidly with increasing depth and transition layer between the mixed layer at the surface and deeper water.

The depth and strength of the thermocline varies from season to season and year to year. It is strongest in the tropics and decrease to non-existent in the polar winter season.

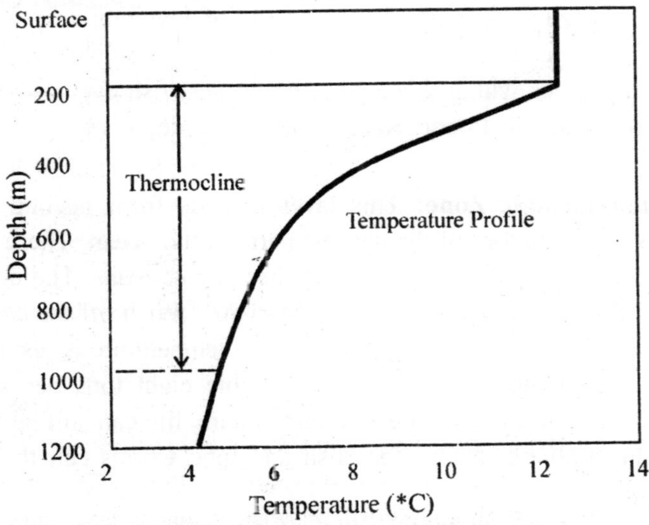

Fig. 6.0.1. Depth. vs. Temperature Profile in Pelagic Zone

The flora found includes surface seaweeds. The fauna consists of many species of fish and some mammals, such as whales and dolphins. Many feed on the abundant plankton.

Mesopelagic Zone: This zone, also known as 'twilight zone' extends from 200 meters to 1000 meters (3281 ft).The light that penetrates to this depth is very faint. In this zone the twinkling lights of bioluminescent creatures become visible. A great diversity of strange and bizarre fishes can be found here.

Bathypelagic Zone: The next layer, also referred to as the midnight zone or the dark zone, extends from 1000 meters down to 4000 meters (13,124 ft). Here the only visible light is that produced by the creatures themselves. The water pressure at this depth is immense, reaching 5,850 pounds per square inch. In spite of the pressure, a surprisingly large number of creatures can be found here. Sperm whales can dive down to this level in search of food. Most of the animals that live at these depths are black or red in color due to the lack of light.

Abyssopelagic Zone: This zone, also known as the abyssal (in Greek means 'no bottom') zone, extends from 4000 meters to 6000 meters (19,686 ft). Abyssal zone has high pressure, cold (around 3° C) water, contains high oxygen and poor nutritional value. The zone supports many species of invertebrates, such as basket stars and tiny squids, and fishes. Three-quarters of the ocean floor lies within this zone. The deepest fish ever discovered was found in the Puerto Rico Trench at a depth of 27,460 feet (8,372 meters).

Hadalpelagic Zone: This layer extends from beyond 6000 meters to the bottom of the deepest parts of the ocean. These areas are mostly found in deep water trenches and canyons. The deepest point in the ocean is located in the Mariana Trench off the coast of Japan at 35,797 feet (10,911 meters). The temperature is just above freezing, and the pressure is an incredible eight tons per square inch. In spite of the pressure and temperature, life can still be found here. Invertebrates such as starfish and tube worms can thrive at these depths.

Mid-ocean ridges (spreading zones between tectonic plates), often with hydrothermal vents, are found in this zone along the ocean floors. Near these vents, chemosynthetic bacteria thrive on the large amounts of hydrogen sulfide and other minerals they emit, which ultimately become fodder for invertebrates and fishes

Ocean and Carbon dioxide Absorption

Ocean removes carbon dioxide from the atmosphere by either absorption or photosynthesis by plants, algae and planktons. Oceans soak up around a quarter to one third of annual CO_2 emissions but should they fail to do so in future, the gas would stay in the atmosphere and could accelerate the greenhouse effect. Reveille's works at Scripps Institute of Oceanography[4] indicate that sea water surface layer has a limited capability of absorbing carbon dioxide, barely one-tenth of calculated buffering capacity. EU research project showed that the North Atlantic, which along with the Antarctic is one of the world's two vital ocean carbon sinks, is absorbing only half the amount of CO_2 as in the mid-1990s.

The more CO_2 the oceans store, the more difficult it will be for them to take up the additional load from the atmosphere and carbon absorption will stagnate even further. Some forms of sea life have suffered from the large amounts of CO_2 absorbed, because of changes in acidity levels, thus affecting biological process of carbon dioxide absorption negatively. Climate change may also affect the biological pump in the future by warming and stratifying the surface ocean, thus reducing the supply of limiting nutrients to surface waters.

Coral reefs

Coral reefs are found in warm shallow waters, as barriers along continents (*e.g.*, the Great Barrier Reef off Australia), fringing islands, and atolls. The dominant organisms in coral reefs are corals. Corals consist of both algae (zooanthellae) and tissues of animal polyp. Since reef waters tend to be nutritionally poor, corals get nutrients through the algae via photosynthesis and also by extending tentacles to obtain plankton from the water. Besides corals, the fauna include several species of microorganisms, invertebrates, fishes, sea urchins, octopuses, and sea stars.

Estuaries

Estuaries are areas where freshwater streams or rivers merge with the ocean. The mixing of waters with different salt concentrations creates a unique ecosystem. Microflora like algae, and macroflora, such as seaweeds, marsh grasses, and mangrove trees (only in the tropics), can be found here. Estuaries support a diverse fauna, including a variety of worms, oysters, crabs, and waterfowl.

Over fishing and pollution have threatened to make oceans into ecological disaster areas. Industrial pollutants that are dumped upstream of estuaries have rendered unsuitable for survival of many marine habitats.

Role of Planktons in Global Warming[6,7,8]

Planktons, inhabitants of fresh water and marine biomes need a special mention, because of their unique properties in relation to global warming. Like plants they are both carbon dioxide absorber and producer of oxygen, a vital element for survival of living organisms. Each year, the North Atlantic ocean announces springtime by producing "blooms" large enough to be seen from space.The latest development in oceanographic remote sensing, however, enables researchers to detect the glow, or phytoplankton fluorescence, from chlorophyll chain[6].

Planktons are primarily divided into broad functional groups[6, 7], as follows:

i. *Phytoplankton* (from Greek *phyton*, or plant), includes different types of algae that live near the water surface where there is sufficient light to support photosynthesis. Among the more important types are the diatoms, cyanobacteria and dinoflagellates.

ii. *Zooplankton* (from Greek *zoon*, or animal), small protozoan or metazoans (*e.g.* crustaceans and other animals) that feed on other plankton and telonemia. Some of the eggs and larvae of larger animals, such as fish, crustaceans, and annelids, are included in this group.

iii. *Bacterioplanktons* are bacteria and archaea, which play an important role in remineralising organic material down the water column.

The planktons that make their own food by photosynthesis are called *phytoplankton* and those small-sized animals that cannot synthesize their own food but feed on organic food available is called *zooplankton*. The phytoplanktons, like the plants on land, are the *primary producers*, and the zooplankton (copepods, salps) that survives on phytoplankton, like the herbivorous animals on land are the *primary consumers*. Carnivorous zooplankton and fish that feed on herbivorous zooplankton are the *secondary consumers*. Bacterioplaktons work as re-mineralising the water required for producing food by photosynthesis. Thus the *plankton as a community* consisting of three broad groups, such as, *producer, consumer and recycler,* has the *capabilities of self sustaining with the help of sunlight.* Planktons are responsible for providing major food source for the sea animals including fishes.

Of the three types, phytoplanktons are the oldest of the known planktons. Scientists consider that the microorganisms originated around three and half billion years ago and can convert water, minerals, and carbon dioxide under sunlight into carbohydrates and oxygen as per following equation:

$$CO_2 + H_2O + \text{nutrients} + \text{light} = \text{organic matter } (C_6 H_{12} O_6) + O_2.$$

The basic requirements for above reaction to occur are as follows:

i. *Light* is required for photosynthesis The top 100-150 m of the ocean surface (photic or euphotic zone) as well as shallow water zones in the coastal areas (fig.6.0) get the required level of light intensity for photosynthesis. These zones are thus favorable for producing phytoplanktons.

ii. *Mineral nutrients* (nitrate, ammonium, phosphate, silicate and iron) are essential for the growth and multiplication of phytoplankton. The nutrient content in the top layer of the ocean is low, hence the above reaction is often limited by the supply of nutrients. However coastal regions are rich in both nutrients and sunlight encouraging the production of planktons. Physical processes such as upwelling (a process in which cooler waters from below are brought up), hydrographic fronts (regions over which a property like temperature or salinity changes sharply in the horizontal), eddies and cyclones are responsible for bringing up nutrients from deeper waters to the surface, thus stimulating primary production of planktons.

iii. *Source of carbon dioxide* for planktons include a) absorption of CO_2 from the atmosphere, b) release through respiration of bacteria and animals, and c) upwelling of CO_2 from dissolved calcium carbonate in shells of benthic animals.

Although large areas of the tropical and sub-tropical oceans have abundant light, they experience relatively low primary production due to the poor availability of nutrients. This is a result of large-scale ocean current and stratification of the water column. In such regions, primary production still usually occurs at greater depth, although at a reduced level (because of reduced light).

Despite significant concentrations of macronutrients, some regions of the ocean are unproductive in the so-called HNLC (high-nutrient, low-chlorophyll) - a term used in marine ecology to describe areas of the ocean where the number of phytoplanktons are low inspite of high macro-nutrient concentrations (nitrate, phosphate, silicic acid). HNLC is thought to be caused by the scarcity of iron (a micro-nutrient which phytoplankton require for photosynthesis) and high grazing rates by zooplanktons that feed on the phytoplankton. The HNLC condition has been observed in the equatorial and sub arctic Pacific Ocean, the Southern Ocean, and in some strong upwelling regimes, such as off central and northern California and off Peru. By adding mineral iron in the iron deficient regions, it is possible to grow phytoplankton. Iron primarily reaches the ocean through the deposition of atmospheric dust on the sea surface. In the famous "iron experiment" in the eastern Pacific and near the South Pole, a solution of iron was spread over a large area to promote phytoplankton growth. An increase in chlorophyll concentration and phytoplankton numbers was noticed.

The oceanic areas adjacent to unproductive or HNLC arid regions of continents have abundant phytoplankton. For example the western Atlantic Ocean areas have plenty of phytoplankton, due to trade winds bringing dust and nutrients from the Sahara Desert in North Africa. This is an example of nature's synchronizing the activities of different biomes for producing phytoplankton.

Phytoplankton plays an important role in the global carbon cycle by regulating atmospheric CO_2. Efforts are being made globally to increase primary production in the world oceans to reduce the atmospheric CO_2. More significantly about 50-75% of the atmosphere's oxygen comes from marine phytoplankton. Besides phytoplankton are also actively involved in the transport of nitrogen, phosphorus, iron and silica in the oceans It has been suggested that large-scale *seeding* of the world's oceans with iron could generate phytoplankton large enough to negate anthropogenic carbon dioxide emission.

3.Deserts

Deserts occur where rainfall is less than 50 cm/year, and cover about one fifth of the Earth's surface. Most deserts occur at low latitudes, such as, Sahara of North Africa and in the southwestern U.S, Mexico, and Australia. Another kind of desert, *cold deserts*, occurs in Utah and Nevada and in parts of western Asia. *Soils often have abundant nutrients because they need only water to become very productive and have little or no organic matter*. Disturbances are common in the form of occasional fires or cold weather, and sudden, infrequent, but intense rains that cause flooding. There are relatively few large mammals in deserts, due to absence of water and shelter from sun and the dominant animals are reptiles.

Desert biomes can be classified according to several characteristics. The four major types of deserts, include, Hot and dry, Semiarid, Coastal and Cold.

Hot and dry desert

The deserts of this type are found in the North America (*e.g.* Great Basin), Southern Asian realm, Neotropical (South and Central America), Ethiopian (Africa) and Australia. The seasons are generally warm throughout the year and very hot in the summer. The winters usually bring little rainfall.

Temperatures exhibit daily extremes because the atmosphere contains little humidity to block the Sun's rays. *Desert surfaces receive a little more than twice the solar radiation received by humid regions and lose almost twice as much heat at night.* The extreme maximum temperature is 49° C, while the minimum drops to -18° C. Evaporation rates regularly exceed rainfall rates. Lowest rainfall is less than 1.5 cm and the highest at 28 cm a year.

Plants are mainly small shrubs and short woody trees with leaves possessing "replete" (fully supported with nutrients) and water-conserving characteristics. In the cacti, the leaves are much-reduced (to spines) and *photosynthetic activity is restricted to the stems*. Some plants open their stomata, openings in leaves that allow for gas exchange, only at night when evaporation rates are lowest.These plants include yuccas, ocotillo, turpentine bush, prickly pears etc.

Semiarid desert

The major deserts of this type include Montana and Great Basin (USA), and the near arctic realm (North America, Newfoundland, Greenland, Russia, Europe and northern Asia).

Winters normally bring low concentrations of rainfall, and average rainfall ranges from 2-4 cm, summer maximum day temperature is 38° C and night temperatures of around 10°C. Cool nights help both plants and animals by reducing moisture loss from transpiration, sweating and breathing. Furthermore, condensation of dew caused by night cooling may equal or exceed the rainfall received by some deserts.

The soil has a fairly low salt concentration with no subsurface water. The spiny plants in this zone with spines shade on the surface significantly reduce transpiration. The silvery or glossy leaves allow them to reflect more radiant energy. Semiarid plants include: creosote bush, white thorn, brittle bushes etc. During the day, insects move around twigs to stay on the shady side, jack rabbits follow the moving shadow of a cactus or shrub. Many animals find protection in underground.

Coastal desert

These deserts occur in moderately cool to warm areas such as the near arctic and Neotropical realm. A good example is the Atacama of Chile. The cool winters of coastal deserts (minimum around -4°C) are followed by moderately long, warm summers (maximum around 35°C). Average rainfall measures 8-13 cm in many areas. The maximum annual precipitation over a long period of years has been 37 cm with a minimum of 5 cm.

The fairly porous soil containing moderate amount of salt and good drainage supports. The plants have thick and fleshy leaves or stems, which can take in large quantities of water when it is available and store it for future use. The list of plants includes the salt bush, black bush, rice grass, and black sage. Some toads seal themselves in burrows with gelatinous secretions for 8-9 months until next rainfall, amphibians with accelerated larval stages thus reaching maturity before the waters evaporate.

Cold desert

These deserts have by cold winters with snowfall and high overall rainfall throughout the winter and occasionally over the summer. They occur in the Antarctic, Greenland and the Nearctic realm. They have short, moist, and moderately warm summers with fairly long, cold winters. The mean winter temperature is between -2 to 4°C and the mean summer temperature is between 21 to 26°C. The winter high snowfall results in annual precipitation ranges from 15 to 46 cm. It contains alluvial soil, which is relatively porous and good drainage leads to leaching out most of the salt. The main plants are deciduous, most having spiny leaves. Widely distributed animals are jack rabbits, kangaroo rats, kangaroo mice, pocket mice etc.

Each kind of desert has its own characteristic plants and animals according to climatic conditions in the region. Variations in the climatic conditions would lead to changes in the ecological cycle and so also the in the type of plants and animals.

Trade winds carrying dust rich in minerals from the Sahara Desert in North Africa have made HNLC or unproductive regions of oceans like western Atlantic Ocean highly productive areas for phytoplankton.

4. Forests

Today, forests occupy approximately one-third of Earth's land area, account for over two-thirds of the leaf area of land plants, and contain about 70% of carbon present in living things. However, forests are becoming major casualties of civilization. The ever growing industrialization and urbanization to cater for the needs of rapidly increasing populations, have resulted in heavy deforestation through occupation of forest lands and consumption of forest resources. The three major types of forest biomes are tropical, temperate and boreal forests (taiga). They are situated at different latitude.

Tropical forest

Tropical forests occurring near equator in an area bounded by 23.5°N & 23.5°S, are characterized by the greatest diversity of species, and have only two seasons, viz., and rainy and dry, with no winter, and almost constant 12 hours daylight. Average temperature is 20-25° C and varies little throughout the year. Precipitation is even throughout the year, with annual rainfall > 2000 mm.

Soil is nutrient-poor and acidic. Decomposition is rapid and soils are subject to heavy leaching. Canopy in tropical forests is multilayered and continuous, allowing little light penetration.

Flora is highly diverse. In 1sq.km, there can be 100 different tree species. Trees are 25-35 m tall, mostly evergreen, with large dark green leaves. Plants such as orchids, bromeliads, vines (lianas), ferns, mosses, and palms are present in tropical forests. Fauna include numerous birds, bats, mammals, and insects.

Further subdivisions of this group are determined by seasonal distribution of rainfall as follows:

- *Evergreen rainforest*: no dry season.
- *Seasonal rainforest*: short dry period in a very wet tropical region. The trees and forest undergo definite seasonal changes. However general character of vegetation remains same as in evergreen rainforest.
- *Semi evergreen forest*: has longer dry season.The upper tree story consists of deciduous trees, while the lower story is still evergreen.
- *Moist/dry deciduous forest (monsoon)*: the length of the dry season increases further as rainfall decreases (all trees are deciduous).

Tropical Rainforests

The intense solar radiation leads to a lot of evaporation, and as warm, moist air rises, it cools, the water condenses, and the water falls back to the earth as rain. Some of the rainforests are known as *cloud forest* or fog *forest*, due to high incidence of low-level cloud cover, usually at the canopy level, such as, Monteverdi cloud forest in Costa Rica.

Tropical rainforest is a major sink for GHGs and producer of oxygen. Rainforests now cover less than 6% of Earth's land surface. Scientists estimate that more than half of all the world's plant and animal species live in tropical rain forests.

Almost all rain forests lie near the equator in three major geographical areas. The list includes, (i) Central America in the Amazon River basin (ii) Africa - Zaire basin, with a small area in West Africa; also eastern Madagascar, and (iii) Indo-Malaysia - west coast of India, Assam, Southeast Asia, New Guinea and Queensland, Australia.

Seventy percent of the plants in the rainforest are trees. The trees in the tropical rainforests produce 40% of Earth's oxygen by photosynthesis through absorption of large amount of carbon dioxide. About 1/4 of all the medicines we use come from rainforest plants. Each of the three largest rainforests has a different group of animal and plant species.

Tropical rain forests, the heartland of ever-growing lush green trees responsible for absorbing large quantities of carbon dioxide producing 40% of the Earth's oxygen have been a major casualty. More than one half of tropical forests have already been destroyed.

Temperate forest

Temperate forests occur in eastern North America, northeastern Asia, and western and central Europe. Temperate zone has well-defined seasons with distinct winter, moderate climate and a growing season of 140-200 days. Temperature varies from -30° C to 30° C. Precipitation (75-150 cm) occurs evenly in a year. Soil is fertile, enriched with decaying litter. Canopy is moderately dense & allows light to penetrate, resulting in well-developed & richly diversified understory vegetation. Flora is characterized by 3-4 tree species per square kilometer. Trees are distinguished by broad leaves that are lost annually and include such species as oak, hickory, beech, hemlock, maple, basswood, cottonwood, elm, willow, and spring-flowering herbs. Fauna is represented by squirrels, rabbits, skunks, birds, deer, mountain lion, bobcat, timber wolf, fox, and black bear.

Further subdivisions of this group are determined by seasonal distribution of rainfall:

- *Moist conifer and evergreen broad-leaved forests*: wet winters & dry summers (rainfall in the winter months and winters are relatively mild).
- *Dry conifer forests*: dominate higher elevation zones; low precipitation.
- *Mediterranean forests*: precipitation in winter, less than 1000 mm per year.
- *Temperate coniferous*: mild winters, high annual precipitation > 2000 mm.

- *Temperate broad-leaved rainforests*: mild, frost-free winters, high precipitation (more than 1500 mm) evenly distributed throughout the year.

Only scattered remnants of original temperate forests remain.

Boreal forest (taiga)

Boreal forests, or taiga, represent the largest terrestrial biome, occurring between 50° & 60°N latitudes. Boreal forests are found in the broad belt of Eurasia and North America: two-thirds in Siberia with the rest in Scandinavia, Alaska, and Canada. Seasons are divided into short, moist, and moderately warm summers and long, cold, and dry winters. The length of the growing season in boreal forests is 130 days. Due to very low temperature, precipitation is mainly 40-100 cm of snow annually. Soil is thin, nutrient-poor, and acidic, canopy permitting low light penetration, & as a result, understory is limited. Flora includes evergreen conifers with needle-like leaves, such as pine, fir. Fauna includes woodpeckers, hawks, bear, fox, wolf, deer, hares, chipmunks; *Current extensive logging in boreal forests may soon cause their disappearance.*

5. Grassland

Grasslands are lands dominated by grasses rather than large shrubs or trees, and of two main types. viz., Tropical grasslands or Savannas and Temperate grasslands

Savanna

Savanna is grassland with scattered individual trees. Savannas of different types cover almost half the surface of Africa (about 5 million square miles, generally central Africa) and large areas of Australia, South America, and India.

Climate making a savanna is warm or hot, and annual rainfall ranging from 50.8 to 127 cm (20-50 inches) per year. Rainfall is concentrated in 6-8 months of the year, followed by long period of drought when fires can occur. *If the rain were well distributed throughout the year, many such areas would become tropical forest.* Three different types of Savannas include, *climatic savannas* (results from climatic conditions), *edaphic savannas* (caused by soil conditions, occur on hills or ridges where soil is shallow, or in valleys in claysoils, waterlogged, in wet-weather) and *derived savanna*, (results from clearing forest land for cultivation). In Africa, a heavy concentration of elephants in protected parkland has created a savanna by destroying trees, thus converting dense woodland into open grassland.

The predominant vegetation consists of grasses and forbs (small broad-leaved plants that grow with grasses). A type of savanna common in Kenya, Tanzania, and Uganda, called grouped-tree grassland, has trees growing only on termite mounds. Frequent fires and large grazing mammals kill seedlings, thus keeping the density of trees and shrubs low. Seasonal fires play a vital role in the savanna's biodiversity. The deep roots remain unharmed by fire and grow when the soil becomes moist. With rains, savanna bunch grasses grow vigorously, to an inch or more in 24 hours. However, controlled burning of far north Australian savannas can result in an overall carbon sink. Other animals (which do not all occur in the same savanna) include giraffes, zebras, buffaloes, kangaroos, mice, moles, snakes, termites, beetles, lions, leopards, hyenas, and elephants. The environmental concerns regarding savannas are mainly due to human activities such as poaching, overgrazing, and clearing of the land for crops.

Temperate grassland

Temperate grasslands have grasses as dominant vegetation (no tree, large shrubs), with large temperatures variation from summer (38° C or 100°F) to winter (-40° C or -40°F). and less rainfall(average 50.8-88.9 cm or 20-35 inches) than temperate grasslands in savannas. Precipitation occurs in the late spring and early summer. Major temperate grasslands include veldts of South Africa, pasta (Hungary), pampas (Argentina & Uruguay), steppes (former Soviet Union), and plains & prairies (Central North America).

As in the savanna, seasonal drought and *occasional fires affect biodiversity but to a lesser extent. Because of nutrient–rich soil, the temperate grasslands are deep and dark, with fertile upper layers.*

Trees, such as cottonwoods, oaks, and willows grow in river valleys, and some nonfood plants, specifically a few hundred species of flowers, grow among the grasses. The fauna in different zones include gazelles, zebras, rhinoceroses, wild horses, lions, wolves, prairie dogs, jack rabbits, deer, etc.

There are also environmental concerns regarding the temperate grasslands. Few natural prairie regions remain because most have been turned into farms or grazing land. This is because they are flat, treeless, covered with grass, and have rich soil. Today, people use steppes to graze livestock and to grow wheat and other crops. Overgrazing, plowing, and excess salts left behind by irrigation waters have harmed some steppes. Strong winds blow loose soil from the ground after plowing, especially during droughts. This causes the dust storms of the Great Plains of the U.S.

6. Tundra

Tundra is the coldest of all the biomes. Tundra comes from the Finnish word *tunturi*, meaning treeless plain. It is a frost-molded landscape with following characteristics:

1. Extremely cold climate & low biotic diversity
2. Simple vegetation structure due to limitation of drainage
3. Short season of growth and reproduction
4. Energy and nutrients in the form of dead organic material
5. Large population oscillations

Tundra is separated into two types:

Arctic Tundra

Arctic tundra is located in the northern hemisphere, encircling the North Pole & extending south to the coniferous forests of the taiga. The arctic is known for its cold, desert-like conditions. The growing season ranges from 50 to 60 days. Yearly precipitation, including melting snow, is 15 to 25 cm (6 to 10 inches). A layer of permanently frozen subsoil called *permafrost* exists, consisting mostly of gravel and finer material. When water saturates the upper surface, bogs and ponds may form, providing moisture for plants. No deep-rooted vegetation forms but a wide variety of plants, about 1,700 kinds, grow the arctic and sub arctic, including: low shrubs, sedges, reindeer mosses, liverworts, & grasses and 400 varieties of flowers. Mammals & birds have additional insulation from fat, breed and raise young quickly in summer, hibernate or migrate (bird) during the winter because food is not abundant.

Recent reports[9] of permafrost (frozen soil in Arctic Tundra) melting is expected to mop up organic matters and bacteria. This may cause release of more CO_2 and CH_4, nitrates and phosphates.While release of GHGs would add to the global warming, nitrates and phosphates would allow novel plants to grow and make the landscape full of domes and pits known as thermokarst – thus changing the tundra's ecology. A combination of melting of ice and thawing permafrost has resulted in submersion of several coastal villages in Alaska.

Alpine Tundra

Alpine tundra is located on mountains throughout the world at high altitude where trees cannot grow. The growing season is approximately 180 days. The night time temperature is usually below freezing. *Unlike the arctic tundra, the soil in the alpine is well drained.* Plants are similar to arctic ones and include: tussock grasses, dwarf trees, small-leafed shrubs, and heaths Animals living in the alpine tundra include mammals: pikas, marmots, mountain goats, sheep, and elk: grouse like birds and insects: beetles, grasshoppers, butterflies.

Carbon & Biodiversity Demarcation Atlas of Biomes[9]

A demonstration atlas correlating carbon and biodiversity has been produced by the World Conservation Monitoring Centre (WCMC) of the UN Environment Programme (UNEP), with support from the German government and the Humane Society International. The atlas is believed to be the first of its kind. Atlas maps those places that contain major species concentrations and where efforts to stop deforestation will produce maximum benefit. The atlas includes regional and national maps for six tropical countries showing where *areas of high carbon storage coincide with areas of biodiversity* importance. It also shows that existing protected areas are high in both carbon and biodiversity.The earth's terrestrial *ecosystems store an estimated 2,000 billion tonnes (Gigatonnes) of carbon (GtC) in the biomass above ground and in the soil*, with a significant proportion of this in the tropics.

The tropical Andes is the richest and most diverse biodiversity hotspot in the world while the Amazon rainforest, the world's largest continuous rainforest area, hosts an estimated quarter of the world's terrestrial species. High biodiversity areas within the tropical *Andes and Amazon account for 11 percent of the total carbon stock* in the area the experts call the *neotropics*. In tropical Africa *over 60 percent of the high biodiversity areas are in high carbon areas and contain a total of 18 billion tonnes of carbon*. Employing the techniques used in the atlas would make it possible to identify where areas of high carbon density and high density of great apes overlap, in order to find where REDD investment could also benefit great ape conservation.

The national maps in the atlas illustrate different ways of identifying areas of biodiversity importance and their overlaps with high carbon areas. In Tanzania, key biodiversity areas contain 17 percent of the country's carbon stock. Vietnam's protected areas cover 32 percent of the land area that has been identified as having high values for both carbon and biodiversity, demonstrating the potential value of the protected area system for meeting both carbon and biodiversity goals. In Papua New Guinea the map illustrates how the centre of the country, which is high in biodiversity, also contains areas of large areas of high carbon stock. It also shows that existing protected areas overlap with only 14 percent of the high carbon areas. 'India's Western Ghats are among the hotspots identified in the atlas.

The prevention of deforestation helps combating climate change, conserving biodiversity from amphibians and birds to primates *Nature has spent millions of years perfecting carbon capture and storage in forests, peatlands, soils and the oceans while evolving the biodiversity that is central to healthy and economically productive ecosystems,* Preservation of biomes is thus an essential step towards minimizing global warming.

Summary

A biome is a large area with similar flora, fauna, and microorganisms. Six biomes covering the Earth's surface include spaces in water (freshwater and marine), forest land (forest and Greenland) and open land (desserts and tundra). Each biome has definite role to play in order to sustain diverse life on the Earth.

The largest biome is that of water occupying two thirds of the Earth's surface. The ocean plays a major role in maintaining the Earth's climatic conditions in different regions, by absorbing carbon dioxide or through photosynthesis by plants, algae and planktons, producing oxygen and equalizing temperature differentials in different zone through ocean currents and upwelling The ocean probably has the richest diversity of species even at less number than species on land. *Oceans soak up around a quarter to one third of annual CO_2 emissions.* Attempts are made to produce planktons in barren areas of ocean by seeding with minerals, like iron. The wanton destruction of such unique sink and biodiversity in the yesteryears must come to an end.

Forests are the second largest biomes responsible for maintaining the climatic conditions by absorbing carbon dioxide and releasing oxygen by photosynthesis. Forests occupy approximately one-third of Earth's land area, account for over two-thirds of the leaf area of land plants, and contain about 70% of carbon present in living things. However forests areas are fast depleting, thus reducing the GHG absorption capacities of another major sink.

On land, biomes are separated primarily by 'latitude'. Terrestrial biomes lying within the 'Arctic Circle' and 'Antarctic Circle' are relatively barren of plant and animal life, and biomes have relatively small amounts of total biomass, smaller energy budgets, can adapt cold and have little effect in changing the climatic condition.

Desert cover one fifth of the Earth's surfaces, receives around twice the solar radiation than humid regions and lose almost same amount of heat at night. Most deserts occur at low latitudes, have few large mammals, and plants limited to small shrubs and short woody trees. In the cacti, photosynthetic *activity is restricted to the stems and not leaves*. Preservation and not destruction of biomes can possibly provide a natural solution to global warming problem.

REFERENCES

1. David D. Kemp, *Exploring Environmental Issues –An integrated approach*, Routledge, London & New York, 2004
2. N.A.Campbell, *Biology*, 4th edition, 1996, The Benjamin/Cummings Publishing Co. Inc., Menlo Park, CA, USA.
3. R.G.Berry and R.J.Chorley, *Atmosphere, Weather, & Climate,* Chapter 2, Solar Radiation and Global Energy Budget, p20-50, Routledge, NY, 2002.
4. Roger Revelle, *The Scripps Institution of Oceanography*, "The Role of the Oceans." *Saturday Review*, 7 May, p. 41,1966. http://www.aip.org/history/climate/co2.htm.
5. J. L.Chapman & M.J. Reiss, *Ecology-Principles and Applications*, 2nd edition, Cambridge University Press, 2000.
6. NASA; The Incredible Glowing Algae; www.nasa.gov/vision/earth/livingthings/glowing_algae.html
7. Plankton: Encyclopedia of Earth. Eds. Cutler J. Cleveland (Washington, D.C.: Environmental Information Coalition, National Council for Science and the Environment). 2008. www.eoearth.org/article/Plankton
8. Plankton: New World Encyclopedia,http://www.newworldencyclopedia. org/entry /Plankton
9. The Economist, August 1st, 2009, p90
10. Carbon & Biodiversity Demarcation Atlas of Biomes: new atlas from the UN Environment Programme /Adobe PDF www.unep.org/pdf/carbon_biodiversity.pdf

8

The Earth Systems-Spheres and Cycles

Earth science generally recognizes four natural spheres, the geosphere (or lithosphere), the hydrosphere, the atmosphere and the biosphere as correspondent to rocks, water, air, and life. The natural cycles for transfer of material and energy amongst these spheres include, rock, carbon, nitrogen and water (hydrological). However increasing human activities in recent decades have affected activities of the natural spheres. The new sphere of human activities has been termed as anthrosphere or man-made sphere. All these spheres are inter-linked with each other in their activities. The impact of anthrospheric activities on other spheres and global warming are to be discussed in this chapter.

The Earth's Spheres- Natural and Man-made[1,2]

The nature's spheres of activities are confined to atmosphere, geosphere, hydrosphere and atmosphere. All the spheres are linked to each other in their activities (tab.7.1). The activities include natural cycles of materials and /or energy transfer amongst the atmosphere, hydrosphere, geosphere, and the biosphere. These various "spheres" act as "reservoirs" that keep materials for different amounts of time, called residence times. The global warming is linked to generation and absorption of carbon, nitrogen and water in their respective cycles by atmosphere, geosphere, hydrosphere and biosphere(tab.7.1) The three material cycles lead to transfer of chemicals from biological to geological systems and are therefore called *biogeochemical cycles*. Processes that affect these transfers are biological processes such as respiration, transpiration, photosynthesis, and decomposition, as well as geological processes such as weathering, soil formation, and sedimentation. However the anthrospheric (or human) activities can affect the natural flows or dynamic cycles of materials and energy. When the rates of anthropogenic disruptions are larger than the balancing capacities of the spheres, the system begins to shift, affecting all levels of the ecosystems through local and global changes.

Geosphere[3]

The geosphere is considered to be the portion of the Earth system that includes *the solid Earth*, its interior, rocks and minerals, landforms and the processes that shape the Earth's surface. The term "lithosphere" is also used instead of geosphere. However the lithosphere only refers to the uppermost layers of the solid Earth (oceanic and continental crust rocks and uppermost mantle). The 94% material of the Earth is comprised of oxygen, iron, silica, and magnesium. The interior of the earth is layered both chemically and mechanically. The geosphere is not static. Its surface (crust) is in a constant state of motion that gives rise to movement of the continents. The unifying theory that explains the continental drift is called *plate tectonics*.

Geosphere is linked to other spheres as follows:

Biosphere: The weathering of the geosphere to form soils provides terrestrial plants with a firm substrate, vital nutrients (phosphorous, nitrogen), and minerals needed for plant growth. In addition, the chemical weathering of the geosphere by water, transports essential nutrients (phosphorous, nitrogen, silica, etc) to the oceans which are used by algae (marine plants) during photosynthesis.

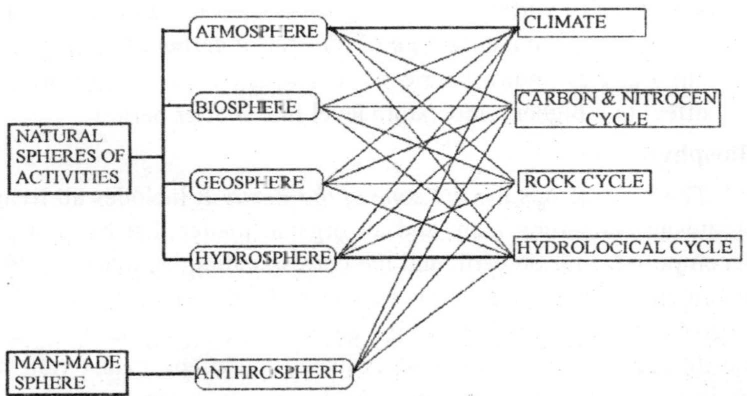

Fig. 7.1 Interlinked Natural Spheres, Cycles & Climate and Anthrosphere

Hydrosphere: The chemical and mechanical erosion by water can cause the rocks to break mechanically into finer particles and can chemically dissolve elements contained in rock-forming minerals, which are carried away downstream to the ocean. The great depth of the Grand Canyon was formed through erosive cutting by the Colorado River. Over long periods of geologic time, erosion by water can wear down mountain ranges (process called peneplanation) and transport this material to be deposited in sedimentary basins.

Atmosphere: Volcanic eruptions inject significant amounts of gases, such as water vapor, carbon dioxide, sulfur dioxide, hydrogen, hydrogen sulfide, etc. into the atmosphere, thus changing the composition and other characteristics of the atmosphere. For example, explosive volcanism that injects large amounts of sulfur dioxide to the upper atmosphere is known to result in cooling of climate in the years immediately following the eruption.

Anthrosphere: Man has been exploiting the mineral resources of the geosphere by mining of raw materials. For example, the mining and subsequent combustion of fossil fuels, and use of limestone for cement making result in the transfer of large quantities of carbon from the geosphere to the atmosphere. By accelerating the natural rate of transfer of carbon dioxide to the atmosphere, anthrospheric activities can enhance greenhouse effect to a higher temperature level in a shorter period.

Biosphere

The biosphere is the *life zone of the Earth.* It includes all living organisms, including man, and all organic matter that has not yet decomposed. Life on earth and the corresponding biosphere readily distinguishes our planet from all others in the solar system. The chemical reactions of life (*e.g.*, photosynthesis-respiration) have transformed the atmosphere from reducing conditions, to an oxidizing environment with free oxygen. The biosphere is structured into a hierarchy known as the *food chain* whereby all life is dependent upon the first tier (*i.e.* mainly the primary producers that are capable of photosynthesis). Energy and mass is transferred from one level of the food chain to the next with an efficiency of about 10%. All organisms are intrinsically linked to their physical environment and the relationship between an organism and its environment is known as the ecosystem. This has been extensively discussed in the previous chapter.

Links of the Biosphere to other components are as follows:

Atmosphere: Life processes involve chemical reactions, such as, photosynthesis and respiration requiring carbon dioxide and oxygen respectively from atmosphere. Other examples of biogenic gases in the atmosphere include methane, dimethylsulfide (DMS), nitrogen, nitrous oxide, ammonia, etc.

Hydrosphere: Water is essential for all living organisms on Earth.Hence the very existence of biosphere depends on water. Water transports the soluble nutrients (phosphate and nitrate) that are needed for plant growth, and the waste products of life's chemical reactions.

Geosphere: The geosphere and biosphere are intimately connected through soils, which consist of mainly mineral matter (alumino-silicate), and also air, organic matter, and water. In fact, one could consider soil as composed of all four spheres (atmosphere, geosphere, biosphere, and hydrosphere). Plant activity, such as root growth and generation of organic acids, are also important for the mechanical and chemical breakdown (weathering) of the geosphere.

Anthrosphere: Human population poses a threat to the biosphere by habitat destruction, especially by the destruction of tropical rainforests (deforestation). This process is driving thousands of species each year to extinction, and these results in fewer sinks to absorb carbon dioxide.

Hydrosphere

The hydrosphere includes *all water on Earth*. Seventy one percent of the Earth is covered by water and only twenty nine percent is land. This unique feature distinguishes our "Blue Planet" from others planets in the solar system, which are devoid of water. It's the heat from the sun which allows water to exist mainly as a liquid, and a major part of this water is contained in the *oceans*. The high heat capacity of this large volume of water (1.35 million cubic kilometers) *buffers the Earth surface from large temperature changes*. However, depending on surface temperatures and pressures water exists also as solid (ice), and gas (water vapor). Water is the universal solvent, and the basis of all life on our Planet. Water is the main agent of chemical and mechanical erosion of the earth surface, which includes breaking rocks, dissolving chemical, forming silts and soils. Hydrologic cycle connects the other spheres with the hydrosphere.

Atmosphere: Water is transferred between the hydrosphere and biosphere by evaporation and precipitation. Energy is also exchanged in this process.

Biosphere: Terrestrial plants withdraw water from the ground using their root systems, and transport water and nutrients through the vascular system to stems and leaves. Evaporation of water from the leaf surface (called transpiration) is effective at transferring water to the atmosphere.

Geosphere: Water is the primary agent for the chemical and mechanical breakdown of rocks, weathering, to form loose rock fragments (regolith) and soil. By the process of erosion, water sculpts the surface of the Earth. Precipitation that falls on the land makes it way by to the sea. The geomorphology of the Earth is unique because of running water.

Anthrosphere: Human activity has significantly impacted the supply and quality of water on Earth through agricultural and industrial practices. Chemical contamination of groundwater, lakes, rivers, and the oceans is threatening the quality of the water supply and affecting biosphere in many parts of the world.

Atmosphere

The atmosphere is the *gaseous envelope that surrounds the Earth.* It begins at the surface of the earth and extends to some 500km above the surface. The lower level, troposphere constitutes the climate system that maintains the conditions suitable for sustaining life and biomes on the earth. The next atmospheric level, stratosphere, contains the ozone layer that protects life on the planet by filtering harmful ultraviolet radiation from the Sun.

Since the Industrial Revolution, the atmospheric heat-trapping "greenhouse" gas contents have consistently increased every year leading to an increase in atmospheric temperatures. This has led to a change in the global climate. Chlorofluorocarbons are effective at depleting the Earth's ozone shield, which protects the earth surface from the harmful effects of ultraviolet radiation.

Atmosphere is linked to other components as follows:

Hydrosphere: The gases of the atmosphere equilibrate with dissolved gases in water through a process known as gas exchange.

Biosphere: The process of photosynthesis, which is a respiration cycle, results in exchanges of carbon dioxide and oxygen between the biosphere and atmosphere.

Geosphere: Volcanic eruptions emit gases to the atmosphere, and atmospheric carbon dioxide dissolves in rainwater to produce a weak acid which is important for the breakdown (weathering) of rock exposed on the surface.

Anthrosphere: Humans breathe air extracting oxygen and emitting carbon dioxide. In addition, our industrial and agricultural activities have changed the chemical composition of the atmosphere.

Anthrosphere

In order to cater to the needs of growing population, we need factories to produce goods and services mostly at the expense of natural resources. The effect of industrialization and the growing population have caused enormous damage to the biosphere, atmosphere, hydrosphere and geosphere. It is essential to control the rate of growth of the anthrosphere, in order to save our planet from destruction. Its links to other components are as follows:

Atmosphere: Industrial and agricultural activities have changed the composition of the atmosphere. For example, the concentration of carbon dioxide in the atmosphere has increased by 26%, and doubled the concentration of methane. The production of chlorofluorocarbons is depleting the earth's ozone layer, our natural defense against ultraviolet radiation.

Hydrosphere: Humans have impacted the hydrosphere by withdrawing large amounts of groundwater for agriculture and by contaminating rivers, lakes, groundwater, and oceans by organic and industrial wastes.

Biosphere: Man has clearly altered the natural biosphere through agricultural activities. A prime example is the slash and burn agricultural practice in the tropics, where rainforest is cut and burned, and the land is converted to pasture. Anthropogenic activities like deforestation and pollution of water resources have led to enormous loss of flora and fauna, both in quantity and diversity.

Geosphere: Mineral and energy resources from the geosphere have fueled the industrial revolution, and that has permitted the human species to increase so prodigiously in number. For example, the exploitation of fossil fuels has increased our standard of living, but an unintended consequence of this action may be climate change and global warming

Cycles

The dynamic natural cycles of actvities amongst the spheres include, the food chain, carbon cycle, nitrogen cycle, water or hydrologic cycle and rock cycle.

1. Food Chain

The food chain was an idea developed by a scientist named Charles Elton in 1927 (ref.3). He described the way plants get energy from sunlight, plant-eating animals get their energy from eating plants, and meat-eating animals get their energy from eating other animals. The idea of a "chain" means that all these animals are linked together, so anything that affects one "link" in the chain affects everything in the chain. The *first link in the chain, the plant, is called the producer, while all the links above it are called consumers. The deforestation leads to disruption of food chain with the disappearance of plants, the only producer.*

2. The Nitrogen Cycle[4,5]

The movement of nitrogen between the atmosphere, biosphere, and geosphere in different forms is described by the nitrogen cycle, one of the major *biogeochemical cycles*. Similar to the carbon cycle, the nitrogen cycle consists of various storage pools of nitrogen and processes by which the pools exchange nitrogen (Fig.7.2).

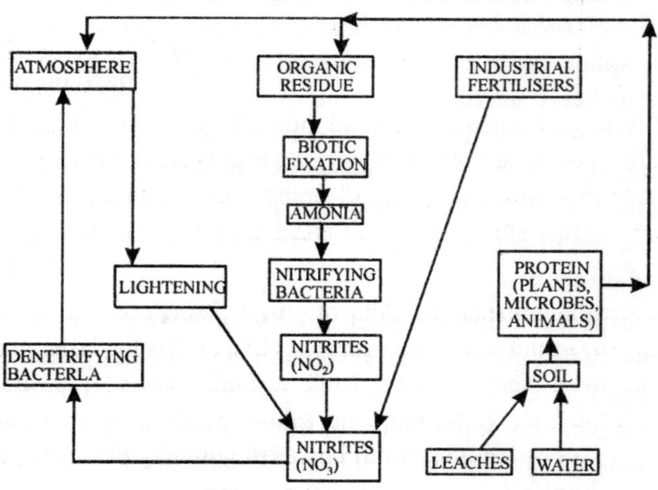

Fig. 7.2 Simplified Diagram of Nitrogen Cycle

The nitrogen cycle is a cycle that mainly exists between microbes and men. All organisms require nitrogen to live and grow.Seventy nine percent of the air is nitrogen. However, due to its inert nature nitrogen present as N_2 molecule cannot be used by plants unless converted or "fixed" in any one of the following usable forms, such as, nitrate ions ($NO_3^{"}$),ammonium(NH_4^+), and organic nitrogen, *e.g.*, urea $(NH_2)_2CO$.

Animals secure their nitrogen and all other nitrogen compounds from plants or animals that have fed on plants.Four processes participate in the cycling of nitrogen through the biosphere:

- Nitrogen Fixation
- Decay
- Nitrification
- Denitrification

Microorganisms play major roles in all four of these processes.

Nitrogen Fixation

The inert nitrogen molecules (N_2) need to be broken into active atoms before they can combine (or get fixed) with other atoms. The breaking process requires the input of substantial amounts of energy. Three processes are responsible for most of the nitrogen fixation include, atmospheric fixation by lightning, biological fixation by certain microbe, and industrial fixation. Atmospheric nitrogen fixation probably contributes some 5–8% of the total nitrogen fixed. In industrial fixation, nitrogen and hydrogen are converted to amonia at high pressure and temperature, which is further processed to fertilisers. Nitrogen-fixing cyanobacteria are essential to maintaining the fertility of semi-aquatic environments like rice paddies. Although the first stable product by nitrogen fixing process is ammonia, this is quickly incorporated into protein and other organic nitrogen compounds. This is achieved either by a host plant, the bacteria itself, or another soil organism by the *nitrogen uptake* process.

Mineralization

After nitrogen is incorporated into organic matter, it is often converted back into inorganic nitrogen by a process called decay or nitrogen mineralization.

$$\text{Organic N} \rightarrow NH_4^+$$

The decomposition of the dead organisms or their excretions by bacteria and fungi leads to the conversion of a significant amount of nitrogen in the dead organism to ammonium. Once in the form of ammonium, nitrogen is available for use by plants or for further transformation into nitrate (NO_3^-) through the process called nitrification.

Nitrification

Ammonia can be taken up directly by plants — usually through their roots. However, most of the ammonia produced by decay is converted into *nitrates,* by a process called nitrification. This is accomplished in two steps:

1. Bacteria of the genus Nitrosomonas oxidize NH_3 to nitrites:

$$NH_4^+ \rightarrow NO_2^-$$

2. Bacteria of genus Nitrobacter oxidize the nitrites to nitrates:

$$NO_2^- \rightarrow NO_3^-$$

Nitrification requires the presence of oxygen, so this process can occur only in oxygen-rich environments like circulating or flowing waters and the surface layers of soils and sediments.

The process of nitrification has some important consequences. The positively charged ammonium ions get attached to negatively charged clay particles and soil organic matter, thus preventing them from being washed out of the soil (or leached) by rainfall. However the negatively charged nitrate ions in soil particles can be washed or leached out, leading to decreased soil fertility and nitrate enrichment of downstream surface and groundwater.

Denitrification

The three processes mentioned above, remove nitrogen from the atmosphere and pass it through ecosystems. Denitrification reduces nitrates to nitrogen gas and to a lesser extent nitrous gas, thus replenishing nitrogen in the atmosphere

$$NO_3 \rightarrow N_2 + N_2O$$

Denitration is an anaerobic process that is carried out by denitrifying bacteria, which convert nitrate to nitrogen in the following sequence:

$$NO_3 \rightarrow NO_2 \rightarrow NO \rightarrow N_2O \rightarrow N_2.$$

Nitric oxide and nitrous oxide are both environmentally important gases. Nitric oxide (NO) contributes to smog, and nitrous oxide (N_2O) is an important GHG, thereby contributing to global climate change.

Once converted to nitrogen by denitrification process, it is unlikely to be reconverted to a biologically available form, because it is a gas and rapidly becomes a part of the atmosphere. Denitrification is the only nitrogen transformation that removes nitrogen from ecosystems (essentially irreversibly), and it roughly balances the amount of nitrogen fixed by the nitrogen fixers described above.

Once again, bacteria are the agents for this process. They live deep in soil and in aquatic sediments where conditions are *anaerobic.* They use nitrates as an alternative to oxygen for the final electron acceptor in their *respiration.* Thus they close the nitrogen cycle.

Human alteration of the Nitrogen cycle and its environmental consequences

The consequences of human-caused nitrogen deposition have affected many aspects of the Earth's ecosystems,such as, precipitation, air and water qualities.

Precipitation: The atmospheric nitrogen oxides form a significant portion of acid rain in the precipitation process.The acid rain has been blamed for forest death and decline in parts of Europe and the Northeast United States. Acid rain can damage and kill aquatic life and vegetation, as well as corrode buildings, bridges, and other structures.

Air Quality: High concentrations of nitrogen oxides in the lower atmosphere are known to damage living tissues, including human lungs, and decrease plant production. Reactive nitrogen (like NO_3^- and NH_4^+) present in surface waters and soils can also enter the atmosphere as the *smog*-component nitric oxide (NO) and the *greenhouse gas* nitrous oxide (N_2O).

Water Quality: Adding large amounts of nitrogen to rivers, lakes, and coastal systems results in *eutrophication*, a condition that occurs in aquatic ecosystems when excessive nutrient concentrations stimulate blooms of harmful algae that deplete oxygen, killing fish and other organisms and ruining water quality. Parts of the Gulf of Mexico, for example, are so inundated with excess fertilizer that the water is clogged with algae, suffocating fish and other marine life.

Effect on Carbon Cycle

The impacts of nitrogen deposition on the global carbon cycle are uncertain, but it is likely that some ecosystems have been fertilized by additional nitrogen, which may boost the capture and storage of carbon. The use of synthetic nitrogen fertilizers that could be added directly to soil has led to an enormous boom in agricultural productivity. But the decrease in natural nitrogen fixation has serious and potentially harmful effects for humans and other organisms because of the disruption of the natural nitrogen cycle.

3. The Hydrologic Cycle[1,2,6,7]

The amount of water on the earth is constant, and is divided up between reservoirs in the oceans, in the air, and on the land. The hydrologic cycle can be thought of as a series of reservoirs, or storage areas, and a set of processes that cause water to move between those reservoirs. Table 7.2 indicates the storage capacities of the reservoirs[7]. The largest reservoir by far is the oceans, which hold about 95% of the earth's water. The remaining 3.5% is the freshwater which is so important to our survival, but about 78% of that is stored in the ice in Antarctica and Greenland. About 21% of freshwater on the earth is groundwater, stored in sediments and rocks below the surface of the earth. The freshwater that we see in rivers, streams, lakes, and rain is less than 1% (0.735%) of the freshwater on the earth, and soil moisture constitutes about 1.5% of water.

Table. 7.2 Water in different reservoirs

Reservoir	*Storage capacity (Total % of Water)*
Ocean	95%
Fresh Water	3.9%
As ice in poles (78%)	2.73%
As groundwater (21%)	0.73%
Soil Moisture	1.5%

The sun is the primary source of energy for all hydro meteorological processes. Although water in the hydrologic cycle is constantly in motion, it never leaves the Earth. The Earth is nearly a "closed system" like a terrarium. This means that the Earth neither gains nor loses much matter, including water. In addition, earth's water is constantly cycling through these reservoirs in a process called the hydrologic cycle. Therefore more water stored in ice sheets means less water in the oceans, and more water in ocean means less ice in artic.

A simplified hydrologic cycle starts with heating caused by solar energy and progresses through stages of evaporation (or sublimation), condensation, precipitation (snow, rain, hail, glaze), groundwater, and runoff (Fig.7.3). Understanding the processes and reservoirs of the hydrologic cycle are fundamental to dealing with many issues, including pollution and global climate change

Unit processes in the cycle[6,7]

The unit processes involve in the hydrological cycle are as follows:

i. *Interception:* Intercepted moisture, stored in the canopy, is the first component of the hydrological cycle to be lost directly back to the atmosphere. The total quantity of intercepted water lost by evaporation can be a significant proportion of the total rainfall. Interception of raindrops by canopies is also a major factor in reducing soil erosion. This has an indirect effect on the hydrological cycle, in that, by conserving surface soil, infiltration is maintained. Precipitation that is not intercepted can be influenced by the following processes;

Stem flow - is the process that directs precipitation down the branches and stems of plants. The redirection of water by this process causes the ground area around the plant's stem to receive additional moisture.In general, deciduous trees have more stem flow than coniferous vegetation.

Canopy drip - some plants have an architecture that directs rainfall or snowfall along the edge of the plant canopy. This is especially true of coniferous vegetation. On the ground, canopy drip creates areas with higher moisture content that are located in a narrow band at the edge of the plant canopy.

Through fall - describes the process of precipitation passing through the plant canopy. This process is controlled by factors like, plant leaf and stem density, type of the precipitation, intensity of the precipitation, and duration of the precipitation event. The amount of precipitation passing through varies greatly with vegetation type.

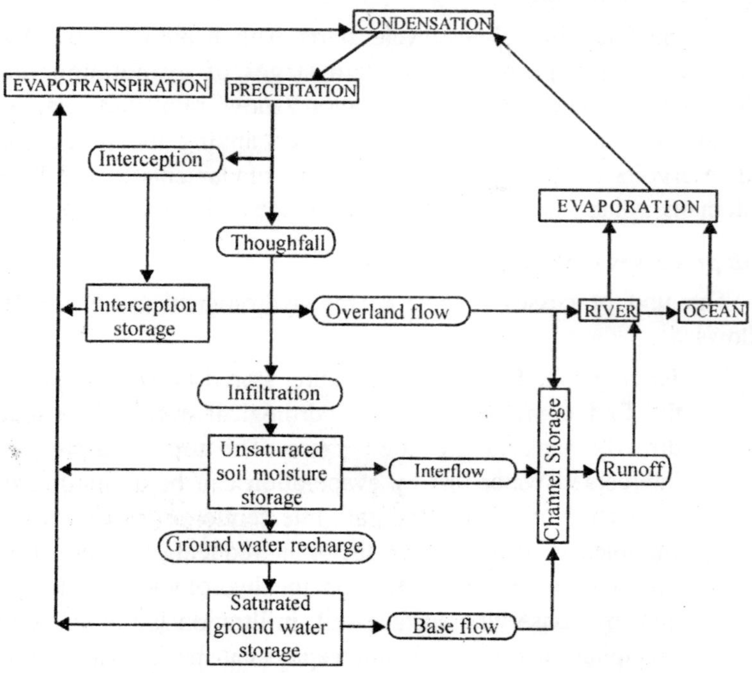

Fig. 7.3. Simplified Diagram of Hydrological Cycle

ii. *Evaporation and Transpiration:* Solar energy evaporates large quantities of water from the oceans and a smaller portion from the land. Plants add some water vapor to the atmosphere through transpiration. Water vapor condenses into clouds and finally precipitates, and most of it falls right back into the oceans, which make up most of this planet.

iii. *Precipitation:* The circulating water vapour in the atmosphere condenses back to liquid water around particulates like dust, called condensation nuclei, with the fall in temperature. The water droplets create clouds, and on collision with other moving clouds form larger droplets, which eventually become heavy enough to precipitate as rain, snow, or hail. Though the amount of precipitation varies widely over the surface of the earth, evaporation and precipitation are globally balanced. In other words, if evaporation increases, precipitation also increases. Rising global temperature is one factor that can cause a worldwide increase in evaporation from the world's oceans, leading to higher overall precipitation.

Some of the clouds are transported over land, where their precipitation falls onto the ground. Most of that precipitation sinks into the shallow layers of soil near the surface (or onto plant surfaces), where it is used immediately by plants, animals, and people. Precipitation comes as rain, snow, sleet, ice pellets, dew, hail, and other types, but the main process is still rain, except in polar regions.

When the water reaches the ground, one of two processes may occur; i) some of the water may evaporate back into the atmosphere, or ii) the water may penetrate the surface and become groundwater. Groundwater either seeps its way to into the oceans, rivers, and streams, or is released back into the atmosphere through transpiration. The balance of water that remains on the earth's surface is runoff, which empties into lakes, rivers and streams and is carried back to the oceans, where the cycle begins again.

iv. *Transpiration:* Plants take up water through their root systems; the water is then pulled up through all parts of the plant and evaporates from the surface of the leaves, a process called *transpiration*. Water that soaks into the soil can also continue to percolate down through the soil profile into *groundwater* reservoirs, called *aquifers*.

v. *Runoff:* When too much rain falls onto the ground to conveniently infiltrate the soil, part of it "runs off". Gravity pulls it directly downhill. This runoff adds greatly to water levels in rivers and lakes. Occasionally, when runoff radically exceeds infiltration, it creates floods and fresh floods.

vi. *Infiltration and Percolation:* Water infiltrates the soil by moving through the *surface*. Percolation is the movement of water through the *soil itself.* Finally, as the water percolates into the deeper layers of the soil, it reaches ground water, which is water below the surface. The upper surface of this underground water is called the "water table". The ground water can intersect with surface streams, it can appear at the surface as springs, and it flows generally downhill toward the ocean.

The Ocean and the Atmosphere

The oceans are the largest reservoir of liquid water, and it's here where most of the evaporation occurs. The amount of water vapor in the atmosphere varies widely over time and from place to place; these variations are reflected as changes in the humidity. Water vapor belongs to GHG family and thus can trap heat in the atmosphere while gases like nitrogen (N_2) and argon (Ar) allow heat to escape to space. The presence of water vapor in the atmosphere helps keep surface air temperatures on the earth comfortable for survival of living organisms and plants, in the range from about -40° C to 55° C. Temperatures on planets without water vapor in the atmosphere, like Mars, stay as low as -100° C.

Since oceans cover around 70% of the earth's surface, most precipitation falls right back into the ocean and the cycle begins again.

Humans and the Hydrologic Cycle

Atmospheric and oceanic circulations are two of the major factors that determine the distribution of climatic zones over the earth. *Changes in the cycle or circulation can result in major climatic shifts*. For example, if average global temperatures continue to increase as they have in recent decades due to anthropogenic increase in GHGs, water that is currently trapped as ice in the polar ice sheets will melt, causing a rise in sea level. Water also expands as it gets warmer, causing further sea level rise.

Many heavily populated coastal areas like New Orleans, Miami, and Bangladesh will be inundated by a meager 1 meter increase in sea level. Additionally, the acceleration of the hydrologic cycle with increasing atmospheric temperature may result in severe weather and extreme conditions. Some scientists believe that the increased frequency and severity of El Niño events in recent decades are due to the acceleration of the hydrologic cycle induced by global warming.

***4. Carbon-Cycle*[8]**

An important ecosystem in connection with global warming is the carbon cycle. Carbon is the basis of all organic molecules.The carbon dioxide is the main source for natural process of heat generation in the atmosphere.

A simplified carbon cycle diagram is shown in Fig.7.4 (ref6).Carbon cycle is the process through which carbon is transferred through the air, ground, plants, animals, and fossil fuels. Large amount of carbon exists in the atmosphere as carbon dioxide (CO_2). Carbon dioxide is cycled by green plants during the process known as photosynthesis to make carbohydrate.The product provides the nourishment of every heterotrophic organism.

The concentration of carbon in living matter (18%) is almost 100 times greater than its concentration in the earth (0.19%). Living matters, therefore extract carbon from the nonliving environment. For life to continue, this carbon must be recycled. Carbon exists in the nonliving environment as:

- carbon dioxide (CO_2) in the atmosphere and dissolved in water (forming HCO_3^-);
- carbonate rocks (limestone and coral = $CaCO_3$);
- deposits of coal, petroleum, and natural gas derived from once-living things or dead organic matter, *e.g.*, humus in the soil.

The global carbon cycle, one of the major biogeochemical cycles, can be divided into geological and biological components. The geological carbon cycle operates on a time scale of millions of years, whereas the biological carbon cycle operates on a time scale of days to thousands of years.

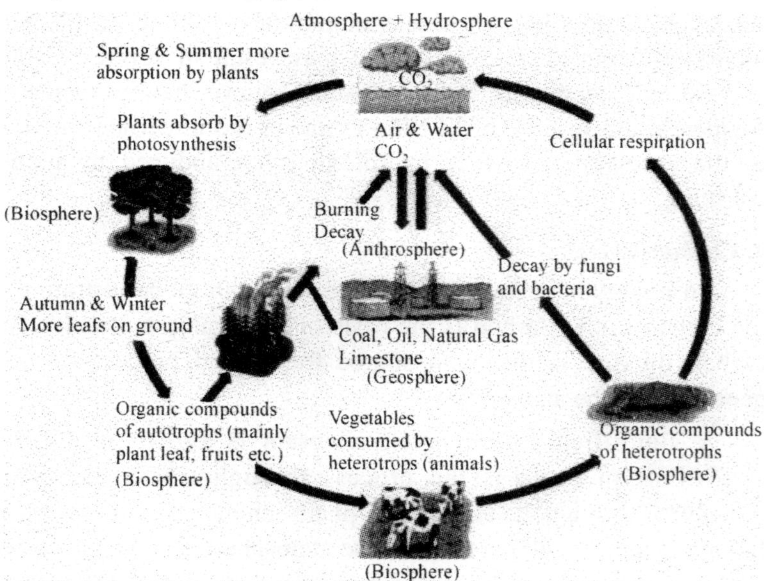

Fig. 7.4 Simplified Diagram of Carbon Cycle

The Geological Carbon Cycle: The geological component of the carbon cycle is where it interacts with the rock cycle in the processes of weathering and dissolution, precipitation of minerals, burial and subduction, and volcanism. In the atmosphere, carbonic acid forms from carbon dioxide (CO_2) and water vapor, reaches the earth as rain, reacts with minerals at the earth's surface(chemical weathering), runs along with surface waters like streams and rivers eventually to the ocean, where they precipitate out as minerals like calcite ($CaCO_3$). The continued deposition and burial, this calcite sediment forms the rock called limestone (***rock cycle***).

This cycle continues as *seafloor spreading* pushes the seafloor under continental margins in the process of *subduction*. As seafloor carbon is pushed deeper into the earth by tectonic *forces*, it *heats* up, eventually melts, and can rise back up to the surface, where it is released as CO_2 and returned to the atmosphere. This return to the atmosphere can occur violently through volcanic eruptions, or more gradually in seeps, vents, and CO_2-rich hotsprings. Tectonic uplift can also expose previously buried limestone. One example of this occurs in the Himalayas where some of the world's highest peaks are formed of material that was once at the bottom of the ocean. Weathering, subduction, and volcanism control atmospheric carbon dioxide concentrations over time periods of hundreds of millions of years.

Carbon Cycle in Biosphere: Biology plays an important role in the movement of carbon between land, ocean, and atmosphere through the processes of photosynthesis and respiration. Carbon enters the biotic world through the action of autotrophs. Autotrophy is the capability of synthesizing organic molecules from inorganic raw materials. Virtually all multicellular life on Earth depends on the (photosynthesis) and the metabolic breakdown (respiration) of those sugars to produce the energy needed for movement, growth, and reproduction.

Photoautotrophs - plants, algae, and some bacteria - use light as the source of the needed energy, through a process called photosynthesis.

Photosynthesis: Green plants produces carbohydrate (sugars) from carbon dioxide and water by sunlight in photosynthesis

Chemoautotrophs — bacteria and archea that do the same but use the energy derived from an oxidation of molecules in their substrate.

Carbon **returns** to the atmosphere and water by biological cycle

Respiration (as CO_2): Respiration by plants and animals and the metabolism result in a *reverse photosynthesis* forming carbon dioxide, which is in turn released to the atmosphere:

$C_6H_{12}O_6$ (organic matter) + $6O_2$ $6CO_2$ + 6 H_2O + energy

Cellular respiration is the process by which organisms release energy from complex organic molecules, typically sugars. All living things, including plants, respire. In the absence of oxygen, anaerobic respiration occurs, producing lactic acid or ethanol.

The majority of the carbon dioxide in the air comes from heterotrophic respiration, and not from the respiration of plants. Other processes of releasing carbon to the atmosphere include the burning & decay of organic bodies, thus producing CO_2 if oxygen is present, otherwise producing methane (CH_4). The amount of carbon taken up by photosynthesis and released back to the atmosphere by respiration each year is about 1,000 times greater than the amount of carbon that moves through the geological cycle on an annual basis.

Carbon cycle in Ocean (Hydro-biosphere)

In the oceans, phytoplankton (microscopic marine plants that form the base of the marine food chain) use carbon to make shells of calcium carbonate ($CaCO_3$). The shells settle to the bottom of the ocean when phytoplankton dies and are buried in the sediments. These and other shells are buried and compressed over time to transform eventually into limestone. Similarly, under certain geological conditions, organic matter can be buried and over time form deposits of carbon-containing fuels, coal and oil. Both limestone and fossil fuel formations are biologically controlled processes and *represent long-term sinks for atmospheric CO_2.*

Uptake and return of CO_2 are not in balance role of anthropogenic carbon. The carbon dioxide content of the atmosphere is gradually and steadily increasing. The increase in CO_2 began with the start of the industrial revolution. and its concentration has risen over 20% by now. This increase is thus of "anthropogenic" origin, that is, caused by human activities: such as, burning fossil fuels (coal, oil, natural gas), and clearing and burning of forests. In addition to global warming & climate change, the increase in CO_2 retards plant growth[9].

Summary

The Earth's natural spheres of activities are known as *geosphere, biosphere, hydrosphere, and atmosphere*. To these natural spheres, a new addition is called the anthrosphere, which covers the areas of human activities. All these spheres are inter-linked to each other in their activities. The cyclic flows or natural cycles of materials, such as, carbon, nitrogen, water and food occur through thesel spheres.

The geosphere is the portion of the Earth system that includes *the solid Earth*, its interior, rocks and minerals, landforms and the processes that shape the Earth's surface. The hydrosphere includes *all water on Earth*. Seventy one percent of the Earth is covered by water and only twenty nine percent is land. The biosphere is the *life zone of the Earth*. It includes all living organisms, including man, and all organic matter that has not yet decomposed. The atmosphere is the *gaseous envelope that surrounds the Earth*..All major natural cycles, such as, carbon, nitrogen and hydrological, are interlinked to all four spheres.

However it is the man-made sphere, known as anthrosphere has been instrumental in causing imbalance to the otherwise dynamically balanced natural cycles through industrial emissions and accumulation of GHGs in the atmosphere.

REFERENCES

1. David D. Kemp, *Exploring Environmental Issues –An integrated approach*, Routledge, London & New York, 2004.
2. J.L.Chapman, and M.J, Reiss, *Ecology, Principles and Applications*, Cambridge University Press, second edition, 1999.
3. C. S. Elton, *Animal Ecology*, 1927. Republished 2001. University of Chicago Press

4. M. Pidwirny, (2006). The Nitrogen Cycle, *Fundamentals of Physical Geography, 2nd Edition.* 2006, http://www.physicalgeography.net/fundamentals/9s.html
5. B. Skinner, *The Dynamic Earth*, page 21. John Wiley & Sons, Inc, 2000
6. Water Framework Directive Centre (WFDC),The Hydrological Cycle, http://www.euwfd.com/html/hydrological_cycle.html
7. IPCC-XXVIII/Doc.13,Conf. on Climate Change, (8.IV.2008),28 Session, Agenda item: 10.Budapest, 9-10 April 2008.
8. The Carbon Cycle, 14 October 2009 **http://users.rcn.com/** jkimball.ma. ultranet/BiologyPages/C/CarbonCycle. html#imbalance.#imbalance
9. Science Daily, Dec6, 2002

Section II: Management

- Managing Global Warming: Methodologies
- Global Efforts –UN and Nations
- Carbon Management System-Emission Limits
- Carbon Sequestration and Geo-engineering
- Fossils and Alternative Energy Sources
- GHG Management in Manufacturing Industries
- Transportation
- Conservation of Biomes and Buffers
- Agriculture and Emission Management
- Waste Management
- Emission Trading and Business
- Corporate Social Responsibility and Climate Change

9

Managing Global Warming: Methodologies

The ecological cycles maintain the concentration ranges of the normally present atmospheric greenhouse gases, through emission and absorption by natural sources and sinks respectively. However, the increase in the emission of greenhouse gases by expanding sources and the decrease in the absorption by sinks have led to monotonous increase in greenhouse gases in the atmosphere.The increase in greenhouse gas concentration at far greater rate than that would occur through natural processes, shall increase the atmospheric temperature at a higher rate, alter the natural carbon and other cycles, and upset the ecological balance.The atmospheric temperature rise would also adversely affect the climate across the globe. The main sources of anthropogenic emissions include the industries, such as, power generation, transport, manufacturing industries (steel, aluminum, cement, chemical, paper & pulp) and agriculture. Anthropogenic activities, such as deforestation, pollution of water resources, destruction of biomes, are mainly responsible for the shrinkage in the capacities to absorb greenhouse gases by the major sinks like forests and ocean.

The management methodologies to negate existing faster rate of global warming include emission control by the industries while improving absorption capacities of the major sinks. Emission control by various industrial sectors include innovative processes to produce more by using less energy and the use of alternative green fuels. The measures to improve sink capacities include prevention of deforestation and destruction of other biomes along with aforestation and preservation of flora and fauna.

Fourth IPCC report in 2007 warns that serious effects of warming have become evident; and the cost of reducing emissions would be far less than the damage they will cause.

Salient features on origin and significance of global warming by GHGs as discussed in earlier chapters are as follows:

i. The *Earth's atmosphere* have different layers and the layer just above the earth's surface, called troposphere dictates weather on the Earth. This layer contains 99% of fixed gases (mainly nitrogen and oxygen) and the rest consists of variable quantities of gases like carbon dioxide and moisture. Variations in the quantities of these two residual gases cause weather change. (Chapter1)

ii. *Weather* is the prevailing condition of the atmosphere with respect to temperature, pressure and precipitation. Long time average of weather over the years is *climate*. An important factor in determining climate is atmospheric temperature. Planetary rotation results in the development of three atmospheric circulation cells in each hemisphere rather than one. These three circulation cells are known as the *Hadley cell, Ferrell cell and Polar cell.* From equator to poles, these three distinct patterns of wind circulation are present at around 30° apart in latitude on either side of the globe. The temperature decreases from equator to poles, and the average temperature in each zone results in different climatic regions. The globe is divided into six climate regions in the three circulatory cells based on average temperature and precipitation values. The increase in average temperature in any region would affect the temperature in other regions due to wind circulation and heat exchange amongst the regions. Therefore, the effect of high GHG emissions in any circulatory cell would be felt in others.

iii. *Solar beams* are responsible for *heating the atmosphere.* Harmful UV-ray of solar beam is filtered by ozone layer in the stratosphere. The heat producing infra-red portion of the solar beam is absorbed by ubiquitous carbon dioxide thus providing the necessary heat to the atmosphere in order to keep the ambient temperature at optimum level. Solar beams produce plant's food by photosynthesis converting absorbed carbon dioxide into carbohydrate, thus reducing atmospheric carbon dioxide content. Thus solar beam while responsible for heating the atmosphere is essential for heat sinks to absorb carbon dioxide by photosynthesis.

iv. *Carbon dioxide*, responsible for heat absorption by the atmosphere has been found to be steadily increasing over the years. The convincing data on the increase of greenhouse gases in the atmosphere were collected for the first time by *Charles David.* Keeling, an US Scientist, who is considered as the *father* of *global warming* issues.

v. Some gases in the atmosphere, such as, carbon dioxide and methane are similar to '*green house*' or enclosed glass vessel, since they allow the visible sunlight to come through, while absorbing heat from the solar beam, and are therefore called greenhouse gases. The different *greenhouse gases* differ in their global warming potentials and life cycles. The higher is the heat absorbing capacity and longer they stay in the atmosphere, greater is the heating effect. The increasing quantities of man-made or anthropogenic GHGs are responsible for global temperature rise.

vi. The world's major communities, classified according to the predominant vegetation and characterized by adaptations of organisms to that particular environment are called biomes. *Biomes* can be grouped in six major types, each of which provides space for different communities of flora and fauna with congenial atmospheric/climatic/soil conditions to their growth. Biomes are responsible for keeping the ecological balance and maintaining of optimum quantities of carbon dioxide in the atmosphere. Hence they need to be preserved.

vii. The four *major spheres of natural activities* on the Earth include *geosphere, biosphere, hydrosphere, and atmosphere.* To these natural spheres, another addition is *anthrosphere* or the areas of human activities. All these areas are inter-linked with each other in their activities. The major activities include natural cycles, such as, carbon, nitrogen, water and food cycles, cover flow of material through these spheres. Oceans are the largest reservoir of water, occupying 70% of the Earth's surface. Water vapor belongs to GHG family and thus its formation leads to trapping of heat by the ocean surface. Evaporation leads to increase humidity and the rain. Rainwater mostly falls on ocean and rest eventually finds its way to ocean.Ocean currents move both vertically and horizontally thus tending to equalize temperature in different zones. Phytoplankton in the ocean water acts as sink for carbon dioxide by photosynthesis.

Nature provides us the ideal atmosphere in which we live in harmony with all other biomes and other spheres. However the activities in man-made sphere, *i.e.*, anthrosphere have affected all other spheres, such as biosphere (de-forestation), geosphere (depletion of mineral resources. degradation of soil quality), hydrosphere (pollution of water by industries) and atmosphere (GHG emissions). In recent times, the increased activities in anthrosphere have greatly affected the natural cycles and activities in the other spheres leading to a large increase in greenhouse gases in the atmosphere. Anthrpogenic GHG emissions, if not managed within a reasonable limit, may significantly change the climate with accompanying hazards.

Management Methodologies

The basic informations on the origin and significance of global warming (as described in the preceding chapters) provide the keys in formulating management strategies & guidelines to control global warming by limiting emissions.

Limit Emission

Management efforts need to be directed towards the reduction of anthropogenic or man-made, enhanced greenhouse gas emission within certain limits. The stipulated emission limits have been set by Kyoto Protocol to the signatories to reduce their collective emissions of six greenhouse gases by at least 5 per cent of 1990 levels by 2012. The signatory countries have set their emission reduction targets and accordingly assigned the emission limits to individual industries and organizations. The limits for industrial sectors are set according to their current emission figures.

The EU figure on CO_2 emissions by various sectors are given in table 8.1[1]. Similar trend in emissions is expected in other industrialized countries. Industries and transports are the major emitters accounting for almost 86% of the total emissions. In EU (group of major industrialized nations & signatories to Kyoto Protocol) major carbon dioxide emitters include the energy sector (34%) and transport (27%) (tab.8.1).These two vital sectors account for more than 61% emissions and therefore need special attention. Other industries account for less than a quarter of total global emissions, although the worst emitters like cement & steel plants belong to this group. The basic trend is similar across the globe. The methodology to be adopted to limit emission in each sector would depend on type and other characteristics of the sector. Fossil fuels are major sources for energy generation as well as emissions. The improvement in energy efficiency is an important factor in controlling emission.

Table.8.1.Sector-wise EU figures of CO_2 emission in 2005 (million tons)

Sectors	*Carbon dioxide emission (m.tons)*	*Percentage of total*
Energy	1562	34
Transport	1247	27
Industry	933	21
Services	285	6.3
Household	478	11

Increase or Preserve Absorption Capacities of Sinks

The nature can possibly cope with, if not fully at least partly, some extra quantities of anthropogenic GHGs in the atmosphere, the amount depending on the available resources for absorbing GHGs in bio-, hydro- and geo- spheres. However human activities have led to depletion of resources in these spheres, thus restricting their capacities to deal with extra GHGs. Along with controlling emissions, it is necessary not only to prevent depletion but also encourage enrichment of resources in other spheres, which are responsible for directly or indirectly acting as sinks. Two most important sinks are the forest and ocean. Each one can absorb enormous quantities of carbon dioxide.The preservation of resources like forests, ocean and associated flora and fauna can lead to absorption of higher quantities of greenhouse gases.

Twin objectives

With the twin objective of reducing emission and enhancing sink capacities, the management techniques require to adopt comprehensive measures, including those stipulated in the Kyoto Protocol and beyond.

We cannot give up our necessities of traveling by car or rail or plane, neither we can give up using electricity, gas etc in our home. For sustaining these and other activities and to cater to the needs of growing population, we need industries to grow, infrastructures to be built and the urbanization to continue.

The best management practice is to find out the ways and means to minimize the greenhouse effect in keeping up with increasing industrial and other activities required by the growing population across the world. The current practices need to be oriented towards savings in energy, while looking for alternate green source of energy.

United Nations Framework Conventions on Climate Change, from Rio to Bali have provided basic norms to minimize greenhouse gas emissions, which the participating nations are expected to follow. Agencies like 'Intergovernmental Panel on Climate Change' (IPCC) have been established in 1988 jointly by the World Meteorological Organization and the United Nations Environmental Programme, to oversee the activities connected with the stipulations made at various conventions at national and international levels.

Developmental activities to counter global warming

While it is the collective responsibility of the corporate, societies and government to reduce energy consumption, encourage afforestation, preserve water resources, and reduce depletion of non-renewable resources, the challenges to scientific communities involve development of viable technologies for *energy conservation* and *reduction of GHG emissions*.

The Kyoto Protocol was enacted in February 2005 and thus legally obligates, all the signatories (mainly industrialized nations), to reduce greenhouse gas emission by 6% of 1990 figure. Japan established NEDO (New Energy and Industrial Technology Development Organization) where extensive work has been carried out on 'global warming countermeasures'[2]. The NEDO report covers voluntary energy conservation action plans for each sector (sectors as per Annex A of Kyoto Protocol) and global warming technology (energy conservation technology and GHG emissions reduction technology). EU sponsored R&D programmes on these areas have also contributed substantially in developing counter-measures for global warming. Countries like USA and Australia, although not signatories to Kyoto Protocol, are also actively pursuing the development work on counter measures to global warming.

Major steps in managing global warming

The few major steps in the management of global warming include the followings:

1. Reduce man-made emissions of greenhouse gases by reducing energy consumption and capping the emitted GHGs. Use alternatives to ozone depleting GHGs.
2. Develop alternate sustainable source of energy based on solar radiation, wind and hydroelectricity, which are completely green and partly green bio-fuel.

3. Conserve forests and actively support aforestation. Forests with its diverse plants and animal is a major sink for carbon dioxide.
4. Conserve water resources, particularly, the sea water constituting 3/4th of the earth, from pollution. Sea water, like forest is a buffer and can absorb large quantity of greenhouse gases.
5. Ensure longer life of equipments & machineries, so as to reduce the carbon footprints of the products. This step also leads to conservation of material and energy by reducing the need for new equipments and machineries.

The methodologies adopted to manage global warming are to be discussed along with the guidelines provided in Kyoto Protocol. The brief outline of the chapters dealing with the management of global warming are as follows:

- Global efforts to formulate strategies on the management of global warming have been discussed in various conventions sponsored by the UN and other international bodies. A major convention is the Kyoto Protocol, which has made significant contributions in the management of global warming (Chapter 9).
- Limiting Carbon Emissions is mandatory to signatories of Kyoto Protocol & voluntary for other nations. It is necessary to follow the standard methods for determination of the quantities of greenhouse gas emissions and the approved methods of reporting in emission reductions. Economic incentive in terms of carbon credit has been the prime mover in reducing emissions as can be seen by the ever growing market for trading on carbon credit (Chapter10).
- Geosequestration of carbon dioxide, by large emitters has been of some help in reducing the release of carbon dioxide to atmosphere. Clean coal technology is also aimed at reducing carbon-dioxide emission (Chapter11).

- Current energy sources are mainly based on fossil fuels. The efforts are being directed towards improving power generation efficiencies of plants using fossil fuels. The hydroelectric power generation is not only the zero-emitter of carbon dioxide, but also based on alternative clean and sustainable resources. Other alternative zero emitting energy sources such as, solar beam, and wind power are yet to be commercialized as large scale alternative to fossil fuels Nuclear power generation does not involve carbon dioxide emission but is not based on sustainable resources. The development and use of biofuels from various sources have been stepped up (Chapter 12).
- Large number of manufacturing and construction industries, such as, mineral products (steel & aluminum), chemical industry, metal production, cement production, with very high emission figures in their efforts to cope with mandatory and or voluntary emission cuts have introduced innovative processes (Chapter 13).
- Transportation industries, including, automobiles, aircrafts, shipping have been active in reducing emission either by making the process more fuel efficient or using alternate fuel. Efforts are being made by automobile manufacturers to reduce GHG emissions per liter of fuel consumed either voluntarily or to conform to mandatory limits set by various countries, such as EU.Research efforts are directed towards reducing transmission losses which can be as high as 40% in automobiles. Aircrafts industry has introduced measures to increase fuel efficiency and also to use partly alternative fuel. Fuel efficient engines have been introduced in the diesel run ships (Chapter 14).
- Biomes are the important sinks for GHGs, esp., forests and oceans.

Forest management is an important subject in the context of global warming in view of fast depleting tropical forests. Carbon atlas shows high carbon storage associated with biodiversity, hence protection of flora and fauna constitutes an important part in preserving forest and ocean resources.

Ocean absorbs both direct heat through evaporation from the surface as well as absorbtion of heat plus carbon dioxide in photosynthesis by phytoplankton. Efforts are being directed towards the use of seeding process for the growth of phytoplankton (Chapter 15).

- Agricultural practices result in GHG emissions through enteric fermentation, manure, rice cultivation, and field burning of agricultural residue. Appropriate methods are available to reduce GHG emissions (Chapter16).
- Management of waste include recycling, reuse and proper treatment & disposal of the waste materials.Recycling of metallic waste reduce carbon footprints of products made, such as stainless steel made from utilizing 70% stainless steel scraps. (Chapter 17)
- Business has grown in the area of carbon project, green marketing, green trading and green projects (*e.g.*, building, power plants) and ecotourism indicating positive growth in the carbon reduction area. Markets for trading carbon directly or indirectly have steadily grown (Chapter18)
- Corporate Social Responsibility is the buzzword today esp., in the area of corporates' involvement in global warming (Chapter19) Current practice of Corporate undertaking voluntary responsibilities for cutting down the emission figures in their corporations would in long run provide the solution pertaining to the problem of global warming without cutting down the GDP.

Summary

Emission management within a specified limit is mandatory to countries (and the various sectors within a country) belonging to Kyoto Protocol. The management of sectorwise emissions are discussed in the subsequent chapters. Brief outlines of management chapters are as follows:

i. Sequestration involves the capture and long-term underground storage of carbon dioxide and use of clean coal to reduce emission to atmosphere.

ii. Power generation from renewable sustainable natural resources, such as, solar beams, water and wind involve zero emission. Also the use of biofuel reduces emission. Improvement in efficiency in running existing fossil fuel based power plants shall considerably reduce emission.

iii. The reduction of carbon emission from transport systems includes weight reduction, more efficient engines and use of alternate fuels.

iv. Various manufacturing sectors, such as, chemical, cement, and steel need to adopt innovative processes to control emission

v. Preservation of sinks, such as biomes, which include oceans and forests with associated biodiversities, is essential to keep up at least the current removal rate, if not better.

v. The waste management, through reuse and recycling, is important in conserving resources and preventing sinks from getting choked.

vi. The corporate needs to be involved in this mission as a part of their social responsibilities

REFERENCES

1. EC 2007 & UIC Energy/CO_2 database: http://www.uic.asso.fr/homepage/ railways& environment_facts &figures .pdf
2. New Energy and Industrial Technology Development Organization (NEDO), Japan. Report on 'global warming countermeasures' entitled as 'Japanese Technologies for Energy Savings/ GHG Emissions Reduction' 2006 Edition, 237 pages.

10

Global Efforts –UN and Nations

The uncontrolled GHG emissions, deforestation, pollution of sea, destruction of biomes, in any region or country can result in global climate change due to global circulation of atmosphere (chapter2), ocean water(chapter 6) and the inter-connecting activities amongst the various spheres (chapter 7).Therefore, there is a need for concerted efforts by the nations across the globe to reduce the menace of GHG emissions in order to minimize global warming.

UN and its agencies have organized several conferences to address this vital issue. The earth summit at Rio (1992) was a good start, but it was the Kyoto Protocol (1997), in which vast majority of participating nations took active part to set the goals, norms and action plans to reduce directly and indirectly GHG emissions. The signatories to Kyoto Protocol gave the commitments for time-bound programme to reduce the emission figures as stipulated in the protocol. Montréal conference was limited to prevent destruction of ozone layer by man-made halocarbons and has been covered in chapter 5. World Energy Conference in Rome (2007), followed by UN-sponsored Bali conference on climate change (2007) –all were directed towards limiting GHG emissions. In the recent one (2009) at Copenhagen, the USA, the opponent to mandatory control on carbon emission has taken major initiative for imposing mandatory carbon capping.

Rio Earth Summit

U N Conference on Environment and Development (UNCED), called as Earth Summit, was held at Rio de Janeiro, on 3-14 June 1992, attended by world leaders. The theme of the conference was 'Environment and sustainable development'. Agenda included, the Rio Declaration on Environment and Development, the Statement of Forest Principles, the United Nations Framework Convention on Climate Change (UNFCCC) and the United Nations Convention on Biological Diversity. Follow-up mechanisms included setting up of Commission on Sustainable Development; Inter-agency Committee on Sustainable Development; High-level Advisory Board on Sustainable Development. In agenda 21, it was decided that *eco-efficiency should be the guiding principle for business and governments alike*. The areas of concerns included the followings.

i. Patterns of production — particularly the production of toxic components, such as lead in gasoline, or poisonous waste — are to be minimized in a systematic manner.
ii. Alternative sources of energy - to replace the use of fossil fuels, linked to global climate change.
iii. Reliance on public transportation systems - to reduce vehicle emissions, congestion in cities
iv. Concern over the growing scarcity of water.

In the Rio Summit it was decided to come up with an agreement that would halt the increasing emission of man-made greenhouse gases into the atmosphere.

Conference of the Parties (COP).

COP comprises of all countries that have ratified the United Nations Framework Convention on Climate Change. COP is responsible for implementing the objectives of the Convention and has been meeting regularly since 1995. More information on outcomes from Conference of the parties (COP) meetings is available at the United Nations Framework Convention on Climate Change (UNFCCC).

Kyoto Protocol

Five years after Rio, the *third session of the Conference of the Parties* to the UN Framework Convention on Climate Change took place in Kyoto, Japan in December 1997, resulting in what is widely known as the Kyoto Protocol. The working agreement of the signatories to Kyoto Protocol, commits developed countries to reduce their collective emissions of six greenhouse gases by at least 5 per cent of 1990 levels by 2012. The Kyoto agreement became legally binding on 16 February 2005 when 132 signatory countries agreed to strive to decrease carbon dioxide emissions. USA, a major emitter was not a signatory of the Kyoto Protocol.

Some of the signatory countries and the corresponding emission cuts are as follows:

i. 15 European Union countries, plus Bulgaria, Czech Republic, Estonia, Lativia, Liechtenstein, Lithonia, Romania, Slovakia, Slovenia and Switzerland: agreed emission cut of 8%.

ii. Although not a signatory, USA agreed for -7 % cut but later dropped

iii. Canada, Hungary, Japan, and Poland: - agreed emission cut of 6 %. In 2006, Canada technically repelled the Kyoto Protocol and allowed emission to continue.

iv. Croatia: agreed emission cut of 5%.

It was a difficult task, since it involves *regulating almost all the activities of modern world* responsible for greenhouse gas emission. A limited amount of emission cut was contemplated by developed countries (excepting USA) while it was impossible to impose similar sanction in the developing economy of the rest of the world.

Although the scope of Kyoto agreement was rather limited, its mechanisms are important because it set standards for further agreements, and created certain important guidelines for tackling the problem, after recognizing the international importance of reversing the steady increase in greenhouse gas emission.

Some of the important features of Kyoto Protocol are as follows:

Article 2: Each party, listed in Annex 1, in achieving its quantified emission limit and reduction commitments (country wise quantified data given in Annexure B), and in order to promote sustainable developments, shall implement policies and measures in accordance with its national circumstances, such as,

i. enhancement of *energy efficiency* in relevant sectors of the national economy. Energy efficiency and conservation of energy in industries are key instruments to reduce GHG emissions.

ii. protection and enhancement of *sinks and reservoirs of greenhouse gases* not controlled by Montreal Protocol. Promotion of sustainable forest management practices, *afforestation and reforestation*, taking into account its commitments under relevant international environment agreement. A major sink and reservoir of main GHG (CO_2) is the forest.

iii. promotion of *sustainable form of agriculture* in light of climate change consideration

iv. R&D and increased use of *new and renewable forms of energy*, of *carbon dioxide sequestration* technologies and of *advanced and innovative environmentally sound technologies.*

v. *progressive reduction* or phasing out of *market imperfections, fiscal incentives, tax* and *duty exemptions* and *subsidies in all GHG emitting sectors* that run counter to the objective of the Convention.

vi. encouragement of appropriate reforms in relevant sectors aimed at *promoting policies and measures which limit or reduce GHGs*, not covered by Montreal Protocol.

vii. measures to limit and/or *reduce emissions of GHGs in transport* sector. In article 2.2 the transport sectors include also *aviation and marine industries*, where GHG emissions reduction need to be worked along with national and international organisations, such as, International Civil Aviation and the International Marine Organization respectively

viii. limitation and /or *reduction of methane emissions* through *recovery and use in waste management,* as well as in the production, transport and distribution of energy.

Article 3

i. ensure that the aggregate anthropogenic *carbon dioxide equivalent emissions* of the GHGs do *not exceed the assigned amounts,* in reducing their overall emissions by at least 5% below 1990 levels in the *commitment period 2008 to 2012.*

ii. may use 1995 as the base level for hydrofluorocarbns, perfluorocarbons, and sulphur hexafluoride, for the purpose (i) as above.

iii. Net changes in greenhouse gas emissions by sources (Annexure A, list of sectors /source) and removal by sinks resulting from direct human-induced land-use change and forestry activities to afforestation, reforestation and deforestation since 1990, measured as verifiable changes in carbon stocks in each commitment period, shall be used to meet the commitments of each party, as included in Annex I. The modalities, rules, and guidelines to calculate changes in carbon stock due to emissions by sources and by sinks in agricultural soils, land-use change and forestry are to be worked out by parties at the earliest.

The aim was to reduce greenhouse gas emissions by an average of 5.2 percent below 1990 levels by 38 nations, who emitted most and who could afford the changeover. In trying to reach greenhouse gas emission standards, many at Kyoto complained that the *European reliance on nuclear power for clean emissions wasn't all that clean.*

India's per capita emission of Carbon Dioxide in 1997 was 1 tonne. China and Brazil had 2 tones of emission each. Hence these countries were not subjected to the Kyoto emission target. With highest end of emission figures of USA (20 tons) and Australia (16 tons), need to participate, otherwise the rest of the world have to shoulder an unfair burden.Although there may be some flaws in Kyoto treaty, till date it's the only international agreement on emission control.

The other important features of the Kyoto Protocol with respect to six GHGs and their emission control in five major sources are covered in Annex A.

1. Annex A

First part contains a list of six greenhouse gases (GHG), such as, Carbon dioxide (CO_2), Methane (CH_4), Nitrous oxide (N_2O), Hydro fluorocarbons (HFCs), Perflurocarbon (PFCs), and Sulphur hexafluoride (SF_6) emissions of which need to be controlled. The changes in emission figures are to be reported in terms of carbon dioxide equivalent emissions (see Article 3i of Kyoto Protocol, also Chapter 5 for C-equivalent).

And the second part of the same annexure lists the GHG emitting sources in five major sectors as follows:

1. *Energy:* Fuel Combustion (Energy industries, manufacturing industries and construction, transport, other sectors, others).Fugitive emissions from fuels (solid fuels, oil and natural gas, other)
2. *Industrial processes:* Mineral Products, Chemical industry, Metal production, other production, production of halocarbons and sulphur hexafluoride, Consumption of halocarbons and SF6.
3. *Solvent and other product use:* Main areas involved are halocarbons.
4. *Agriculture:* Topic includes, enteric fermentation, manure management, rice cultivation, agricultural soils, prescribed burning of savannas, field burning of agricultural residues, other.
5. *Waste:* This topic includes solid waste disposal on land, material recycling, waste water handling, waste incineration etc.

Penalties are to be imposed for signatory countries not reaching the target by 2012. By this time a second set of commitments up to 2020 will be negotiated. Countries failed to reach first target have it added to second target plus a 30per cent penalty for the shortfall,

November-07-Rome- World Energy Conference

The biggest conference on the subject brings together world leaders and corporate executives once in every five years. The conference has expressed concern of climate changes and its potentially disastrous economic and social effects with the use of fossil fuels.

European Commission president, in his keynote address said that Europe would lead the way toward conservation and the efficient use of energy in order to mitigate worst effects of climate change. He has urged the developing countries, like India and China to find a way to economic growth while still regulating emissions. This issue of involving developing countries in the manadatory regime of emission control has been repeated again in the Bali conference(2007) and more recent Copenhagen meet (2009). The last two conferences were held to replace Kyoto Protocol expiring in 2012.

Meanwhile, the EC plans to take a bold step in framing legislative proposals to cut automobile emissions to 120g of carbon dioxide per Km by 2012, a standard that only 8% vehicles in Europe are able to meet in 2006.

Dec-07 Bali-Climate Change Conference

Bali Conference was held in december, 2007, in order to replace Kyoto Protocol on its expiry in 2012.The conference was attended, amongst other dignitaries, by UN secretary general, Ban Ki-moon. Unlike Kyoto, in Bali conference, US delegates agreed "to go forward and join consensus"[1]. Bringing USA in the mission to curb carbon emission was a major achievement at Bali. Excepting for a statement of good intentions, Bali conference did not produced anything concrete to replace Kyoto protocol, with a new international agreement by 2012.

The 'reduction in carbon emissions from deforestation and forest degradation' was excluded from Kyoto due partly on scientific grounds as to the 'precise effects ' and also due to non-availability of a method to measure 'carbon loss'. However mounting evidence of the destruction of tropical forests causing as much as 20% of GHG emissions, along with better technologies for measurements have led to lengthy discussion on this subject, including 'avoidable deforestation' in post 2012 strategy (when Kyoto Protocol expires) at Bali conference[2].

Bali conference stressed the urgent need for "meaningful action to reduce emissions from deforestation and forest degradation". A concrete step in that direction is to have a "work programme" to test various approaches.

Under the Kyoto Protocol, expiring in 2012, rewards for storing carbon through trees are only given for reforestation or planting new forests. This restriction has become unjustifiable in view of the facts that destruction of tropical forests resulted in shrinkage of storing capacities equivalent to 20% of GHGs emissions.

If deforestation is taken into account then Indonesia & Brazil soar up to become world's 3rd and 4th largest emitters. However, Brazil leaders want a system of reward for countries bringing their deforestation rates down from pre-1990 levels, and keeping them there. If so happen then Brazil may be only nation that qualifies.

The decisions taken at Bali call for a couple of more years of trial and error to find the best ways to deal with this subject. Investors in carbon markets will be first to shun these conservation projects that are manifestly failing because of bad government or corruption[2].

Post Bali efforts

Despite not reaching to any broad consensus at Bali, efforts are being made by individual states including US to cut down carbon emission. For example,

- EU imposing a stricter carbon limit in automobile emissions,
- China imposing fuel-economy regulations,
- America's Congress approving a bill on fuel economy, by 2020,

- California will require all petrol sold there to meet low-carbon regulations
- In 2008, environment ministers of G-8 (group of eight industrialized nations)including China & India began talks to pave the way for forthcoming G-8 summit.

These are indicatives of the major economies trying to set the positive trends towards reducing carbon emission.

Indonesia, the host to Bali conference, with a population of 235 million and one of the largest carbon footprints outside of developed countries, has outlined a plan to slash greenhouse gas emissions by 17% in 2025. Indonesia hope to achieve the target by reductions in forest burning and cutting the oil consumption. Mr. Boer, head, UN Climate Change Secretariat, urged USA to follow Indonesia rather than President Bush's promised capping by 2025, which is 'not enough' and paltry compared to Indonesia.[3]

Table 9.1 for Climate Performance Index of 10 Best & Worst

BEST	SCORE	WORST	SCORE
1 Sweden	65.6	56 Saudi Arabia	30.0
2 Germany	64.5	55 US	33.4
2 Iceland	62.6	54 Australia	35.5
4 Mexico	62.5	53 Canada	37.6
5 India	62.4	52 Luxembourg	39.2
6 Hungary	61.0	51 South Korea	41.3
7 UK	59.2	50 Russia	43.9
8 Brazil	59.0	49 Malaysia	44.2
9 Switzerland	59.0	48 Kazakhstan	44.6
10 Argentina	58.5	47 Ukraine	44.7

Source: Bali conference Report, Hindustan Times, Dec 9, 2007, Mumbai, India

Climate Change Performance Index

The results of a study on the climate change performance of 56 countries reported at Bali Conference. The score is based on per capita emission trends, absolute energy-related CO_2 emissions of a country and on domestic and international climate policy of a country. The scores of best ten and the worst ten countries are indicated in the tab.9.1. In the list for top ten performers (tab.9.1), India occupies the 4th position, with the performance index of 62.4, which is nearly equal to that of top performer, viz, Sweden with a performance index of 65.6. However the per capita emission in India would be low in comparison to that of Sweden, due to vast difference in population and the scale of developmen

UN Framework Convention on Climate Change (UNFCCC), Dec1-12, 2008, Poznan, Poland.

Discussions in UNFCCC at Poland were centred around following areas:

- Moving towards a 'negotiating text' for a treaty to combat climate change that would be finalised by the next summit in Copenhagen, December, 2009.
- On the operationalisation of the Adaptation Fund meant to help least developed countries (LDC) to cope with climate change, it remains unresolved.
- On assessing risks due to climate change and the 'possibility of creating an insurance mechanism' to cover these risks.
- On reforming the clean development mechanism (CDM) that rewards green technologies in developing countries. There was still disagreement on whether carbon capture and storage (mainly emissions from coal-fired power plants) would be included in it wholly, partially, now, later or not at all.
- On rewarding developing countries that control deforestation, there was no agreement on how to measure this benefit in the form of reduced greenhouse gas emissions.

Rating Countries based on per capita Emissions

A better index to emissions of anthropogenic or man-made gases by different countries would be per capita emissions rather than the absolute value. The per capita emission is a logically correct index of performance with repect to emission reduction by nations across the globe. However, from global warming point of view, the absolute emission figure is the one which would determine the quantity of greenhouse gas in the atmosphere and thus the global warming potential (GWP).

The world's countries contribute different amounts of heat-trapping gases to the atmosphere. The table 9.2 shows data compiled by the Energy Information Agency (Department of Energy, USA), which estimates carbon dioxide emissions from all sources of fossil fuel burning and consumption. The list of the 20 countries with the highest carbon dioxide emissions (data are for 2006) is included in table 9.2. Australia tops the list followed by US, Canada, & Saudi Arabia.Another group of developed countries with next level(around 10) includes UK, S.Korea, S.Africa, Japan, Germany and Spain. Lowest per capita emission figure is that of India, which is 1.16.

Table 9.2 Emissions & Per Capita Data for Top 20 Nations

	Country	*Total Emissions (million tons CO_2)*	*Per Capita Emissions (tons/capita)*
1.	China	6017.69	4.58
2.	United States	5902.75	19.78
3.	Russia	1704.36	12.00
4.	India	1293.17	1.16
5.	Japan	1246.76	9.78
6.	Germany	857.60	10.40
7.	Canada	614.33	18.81
8.	United Kingdom	585.71	9.66
9.	South Korea	514.53	10.53
10.	Iran	471.48	7.25
11.	Italy	468.19	8.05
12.	South Africa	443.58	10.04
13.	Mexico	435.60	4.05
14.	Saudi Arabia	424.08	15.70
15.	France	417.75	6.60
16.	Australia	417.06	20.58
17.	Brazil	377.24	2.01
18.	Spain	372.61	9.22
19.	Ukraine	328.72	7.05
20.	Poland	303.42	7.87

The industrialised countries' earlier commitments of financing and technology transfers to developing countries to help them combat climate change are now withdrawn. Industrialised countries are pressing India and China to make legally binding commitments to cap their greenhouse gas emissions, though per capita emissions in India are just over one tonne of carbon dioxide a year, compared to 11 tonnes in European Union (EU) countries and 20 in the US.

Developing countries are struggling to find money for Reducing Emissions from Deforestation and Forest Degradation (REDD) in a post-2012 climate deal. If they do manage to find a significant sum of money, the Carbon and Biodiversity Demonstration Atlas produced by the World Conservation Monitoring Centre (WCMC) of the UN Environment Programme (UNEP) will come in very handy indeed (see Chapters 6 & 15).

Carbon Emission and GDP

The ***carbon intensity***, also called per capita annual emissions, is a measure of how much carbon equivalents (CO_2e) are emitted per capita of GDP. The inverse is the metric called ***carbon productivity*** which is the amount of GDP product per unit of carbon equivalent. Carbon intensity is thus the amount of carbon dioxide emitted per unit of economic growth (commonly 1000 dollar GDP)

According to figures published by the United States Department of Energy, China in 2006 emitted 2.85 tonnes of carbon dioxide from fossil fuels for every $1,000 (604 pounds) of gross domestic product (GDP), around 15 percent lower than a decade earlier. In comparison, the United States in 2006 emitted 0.52 tonnes of carbon dioxide for every $1,000 of GDP, while Switzerland produced 0.17 tonnes, and impoverished Chad just 0.07 tonnes. The corresponding figures for Russia, S.Africa, and India are 4.5, 2.42, and 1.80 tonnes respectively.

The India's stand that the current climate change negotiations under the auspices of UN Framework Convention on Climate Change (UNFCC) are being skewed in favor of the industrialized nations got another shot in the arm. *Purported pre-release of a McKinsey report projects that India will continue to be one of the LEAST*

Carbon Intensive countries in the world despite an economic growth rate of 7.5%. This second endorsement follows the recent report by the World Bank saying that India is right in resisting the mandatory emissions reduction[4]. China has pledged to cut the amount of carbon dioxide produced for each unit of economic growth by 40-45 percent by 2020, compared with 2005 levels.On same account, India's figure is 25 percent by 2020 from 2005 levels[5] Copenhagen Climate Conference November-December, 2009.

The recent United Nation's Climate Change Conference held was to replace Kyoto Protocol (setting targets for cutting emissions) by a new deal. The aim of the conference was to have a comprehensive political agreement that puts the countries on a clear path to concluding a binding agreement in 2010.The interim agreement should lead to immediate action and a broad road map for future treaties, including:

1. Political commitments for mid-term action by all major countries; economy-wide emission reduction targets for developed countries, and quantified reduction mitigation actions by major developing countries, including India.
2. A 'prompt start' on adoption, forestry, technology and capacity building activities and support in developing countries.
3. The legally binding agreements are to be finalized over the coming year, including: a framework for verifiable mitigation commitments by all major countries; new agreements for sustained mitigation and adaptation support to developing countries; and a system to verify countries' actions and supports.
4. A clear mandate to conclude negotiations on a legally binding agreement at COP 16 in December 2010.

The USA, a non-signatory to Kyoto Protocol, has for the first time opted for mandatory emission cut at the Copenhagen meet. Developing countries like India and China are against mandatory regime. They agree to impose voluntary emission cut.

The negotiotations are proceeding on parallel track under the UN framework Convention on Climate Change (UNFCCC), with US participation, and also under UNFCCC's Kyoto Protocol, without the USA.

Framework for Mitigation Commitments

The final agreement should clearly define the nature of mitigation commitments and how they are to be reflected in a final agreement. Following the principle s of 'common but differentiated responsibilities', it should allow varying forms and levels of commitments depending on national circumstance:

- Absolute economy-wide emission targets for all developed countries, and
- A wider range of quantifiable policy-based commitments for major developing countries, like India & China (*e.g.*, sectoral emission targets, energy efficiency standards, renewable energy targets, sustainable forestry goals).

The agreement should launch and support a process, such as a "registry" process, to elaborate country-specific commitments for the major developing countries and to align support for them. It also should go as far as possible in defining implementation and accounting rules.

Support for Developing Countries. The agreement should broadly establish the mechanisms, sources, and levels of support to be provided in a final agreement for adaptation, capacity building, forestry and technology deployment in developing countries. It should: set initial funding levels and a timetable for periodic replenishment; set criteria to determine countries' contributions to and/or eligibility for support; rely on, rather than replicate.

A Sound System of Verification. The agreement should establish basic terms for the measurement, reporting and verification of countries' mitigation actions, and of support for developing country efforts, as called for in the Bali Action Plan. Building on existing reporting and review requirements under the UNFCCC and Kyoto

Protocol, it should require annual emissions inventories by all major-emitting countries (with a phase-in period and support for developing countries); national verification of countries' mitigation commitments; and, regular implementation reports subject to international review.[6]

Opposition to Copenhagen Protocol

The major developing nations like Brazil, South Africa, India and China (BASIC) are against imposition of mandatory sanctions on emissions on developing economies, without finanacial aid to adopt clean technologies to combat emissions.Although these countries are also major emitters, particularly China followed by India, however they have very low per capita emission figure in comparison with advanced countries. In India, 50% of the villages have no electricity. There are no alternative than to have coal based big power plants to bridge the large gap in the basic needs of a huge population, in an appropriate timeframe. Under these circumstances, India cannot afford to impose sanctions on the emitting industries till these basic amenities are extended to cover the wide spectrum of the population. Hence legal binding on emission should be accompanied by financial assistance to developing nations in adopting green technologies.

Post Copenhagen Meet of BASIC

The BASIC meeting at Dehli in January, 2010 follows the Copenhagen Climate Change Conference where it was decided to continue negotiations in two tracks relating to the Bali Action Plan and the Kyoto Protocol for another one year so as to have a final outcome at the sixteenth Conference of Parties at Mexico in 2010. The meeting called for finalising a legally binding treaty on reduction of carbon emission latest by 2011 and indicated that the world could not wait indefinitely. A legally binding outcome should be concluded during climate change meet at Cancun, Mexico in 2010, or at the latest in South Africa by November, 2011.The news reports that domestic legislation in the US had been postponed has been a major concern to reach an internationally legally binding agreement at an early date.

EU to lead

EU led the fight against climate change through its ground-braking Emissions Trading Scheme in accordance with Kyoto Protocol and the new emission proposal of "20/20/20 by 2020" in which the emission cut would be 20% below 1990 level, plus a 20% gain in energy efficiency, plus 20% of energy from the renewables.This plan needs the approval of Council of ministers and the European Parliament. It is meeting hefty opposition from heavy industries and coal-dependent countries like Poland.[7]

2010 United Nations Climate Change Conference

The conference was held in Cancun, Mexico, 29 Nov. to 10 Dec. 2010. The conference is officially referred to as the 16th session of the Conference of the Parties (COP 16) to the United Nations Framework Convention on Climate Change (UNFCCC) and the 6th session of the Conference of the Parties serving as the meeting of the Parties (CMP 6) to the Kyoto Protocol. In addition, the two permanent subsidiary bodies of the UNFCCC – the Subsidiary Body for Scientific and Technological Advice (SBSTA) and the Subsidiary Body for Implementation (SBI) – held their 33rd sessions. The 2009 United Nations Climate Change Conference at Copenhagen extended the mandates of the two temporary subsidiary bodies, the Ad Hoc Working Group on Further Commitments for Annex I Parties under the Kyoto Protocol (AWG-KP) and the Ad Hoc Working Group on Long-term Cooperative Action under the Convention (AWG-LCA), and they met as well.

The outcome of the summit was an agreement, though not a binding treaty, to limit global warming to less than 2° C above pre-industrial levels and calls on rich countries to reduce their greenhouse gas emissions as pledged in the Copenhagen Accord, and for developing countries to plan to reduce their emissions. The agreement includes a "Green Climate Fund," proposed to be worth $100 billion a year by 2020, to assist poorer countries in financing emission reductions and adaptation. However there was no agreement on how to extend the Kyoto Protocol, or how the $100 billion a year for the Green Climate Fund will be raised, or whether developing countries should have binding emissions reductions or whether rich countries would have to reduce emissions first. To most delegates, though they approved it, the agreement "*fell woefully short of action needed*"[8].

Summary

i. The third Conference of the Parties to the UN Framework Convention on Climate Change held in Kyoto, Japan, resulting in what is widely known as the Kyoto Protocol. This working agreement of the signatories commits developed countries to reduce their collective emissions of six greenhouse gases by at least 5 per cent of 1990 levels by 2012. The Kyoto agreement became legally binding on 16 February 2005 when 132 signatory countries agreed to strive to decrease carbon dioxide emissions.

ii. Kyoto Protocol is the most effective agreement on climate change. Most of the leading signatory nations have taken effective steps as stipulated in the protocol to minimize emission.

iii. Bali Conference of nations mainly dealt with deforestation but there was no agreement reached amongst the participating nations.

iv. The hectic activities amongst the nations would hopefully result in adopting some concrete steps to curb emissions as planned in the Copenhagen conference.

REFERENCES

1. *The Economist,* December 27, 2007, p. 10.
2. *The Economist*, p. 98, Dec 22, 07-Jan 4th 08.
3. *The Statesman*, Kolkata, India 26 May, 2008, p. 3.
4. Dr.Vandana Prakash:India one of the least Carbon Intensive Countries in the World; McKinsey Reports, Published on May 24th, 2009. http:// eco worldly.com/2009/05/24/india-one-of-least-carbon-intensive-countries-in-the-world-mckinsey-reports

5. *Times of India*, 28th Nov'09
6. A Copenhagen Climate Agreement; permission to use the material in the website has been obtained from Tom Steinfeldt. Pew Center on Global Climate Change; 11May, 2010.; http://www.pewclimate.org/international/copenhagen-climate-agreement,
7. *The Environment*, a review in The Economist, special issue, The World in 2009, special section on 'The environment', p103
8. Doyle, Alister, Analysis Climate talks:18 years, too little action. *Reuters,* 17 Dec'2010, Retrieved 2011-01-8.

11

Carbon Management System-Emission Limits

The carbon dioxide concentrations in the atmosphere are naturally regulated by processes such as, the "carbon cycle" (see fig in chapter 6), which includes natural movement ("flux") of carbon between atmosphere, land, ocean, and forests.

The net 6.1 billion metric tons of anthropogenic (man-made) carbon dioxide emissions are produced each year (measured in terms of carbon dioxide equivalent). An estimated 3.2 billion metric tons of carbon is added to the atmosphere annually[1,2]. Nearly fifty percent of the emissions are absorbed by natural process. Rest stays in the atmosphere for long period of around 100 years or more, depending on the type of gas. The cumulative effect can have disastrous consequences. While anthropogenic emissions are increasing due to higher economic growth, the sinks for carbon dioxide emission are shrinking fast because of deforestation and degradation of water resources of oceans around the world. The increase in the gap between emissions and absorption has resulted in a net increase of GHGs in the atmosphere. Anthropogenic GHGs emissions in the atmosphere need to be regulated in order to minimize global warming.

Carbon dioxide, the main culprit, accounts for 64% of the current total greenhouse gas emissions. Methane, originating from landfills, coal mines, oil & gas operations, and agriculture represents 9 percent of total emissions Nitrous oxide (5 percent of total emissions) is emitted from burning fossil fuels and through the use of certain fertilizers and industrial processes. Other human-made gases (2 percent of total emissions) are released as byproducts of industrial processes and through leakage.

In reporting total emission figure of all the greenhouse gases,the emissions from non-carbon dioxide gases are converted to carbon dioxide equivalent with respect to global warming potentials. The total emission is the summation of carbon dioxide emission and the carbon dioxide equivalents of other GHGs. The figure is simply termed as 'carbon' emission. A major percentage of the emission is due to energy consumption Energy consumption is directly related to emission and can be converted to carbon emission units. The improvement in the carbon absorption capacity of the sink (*e.g.* aforestation) is also entitled for carbon credit. The cap and trade carbon management system for limiting both direct and indirect anthropogenic greenhouse gas emissions is based on Kyoto Protocol.

The signatories to Kyoto Protocol have agreed to cut down emission of at least 5 per cent of 1990 figure by 2012. The Kyoto agreement became legally binding on 16 February 2005 when 132 signatory countries agreed to strive to decrease carbon dioxide emissions. Accordingly the industries in those countries are assigned emission limits in terms of carbon credits. Industries need to compensate for exceeding the limits by buying the required carbon credit from mandatory market, and those with surplus credits can sell the same. The corporations in other non-Kyoto Protocol countries can also buy and sale carbon credits in the voluntary markets. Carbon management is a mandatory requirement by industries belonging to Kyoto Protocol countries. As a part of CSR activities (chapter 19), voluntary management of carbon by corporations around the world has become an essential ingredient of current business practice.

Assessment of Emission & Emission Reduction

In the carbon management process it is essaential to determine emission data on the following lines:

i. Determination of quantum of GHG emissions from the use of fossil fuel and conversion of the same to carbon equivalent.

ii. Determination of Energy consumption; Energy consumption is directly related to emission and need to be converted to carbon emission units.

iii. Footprint assessment: - a full carbon footprint of a product incorporates every stage of the product's life, from cradle to grave. The sum total of carbon foot prints of all inputs, processing steps, quality control, packaging and transportation would determine total carbon footprint of the product.

iv. Carbon offset management requires data on carbon reduction in the offset methods, such as, use of alternate non-fossil fuels, aforestation etc.

v. The carbon emission figures are required to be expressed in different units by various sectors, such as, million tons by industries, gms per km by automobiles etc.Emission reduction units are to be expressed in different units as required by various mechanisms for transfer under Kyoto Protocol & others.

The sum total of (i) and (ii) is the total carbon emission from the concerned organisation. Carbon offset figure (iv) can be deducted from total emission data to arrive at net carbon emission.The carbon footprint of the product indicates the carbon emission in producing the product. The lower carbon footprint figure indicates lower carbon emission.

i. Amount of GHG emissions and carbon equivalent

Carbon emissions from industries can be determined by following the procedures as stipulated in various standards.Some important standards are as follows:

a. *ASTM D6866 for CO_2 & Methane:* The prescribed method in ASTM (American Standard for Testing Materials) D6866 can be used to attest the percentage of renewable CO_2

being emitted into the atmosphere or the amount of methane being combusted. This is of particular importance for industries such as the electricity producers that use biomass material or landfill gas and monetary agencies needing verification of fossil CO_2 emissions. Also useful for industries where the amount of renewable content is in doubt. One of the industries that benefits from the use of ASTM D6866 is the Waste-to-Energy (Energy-from-Waste) industry. This method also applies to industries such as electrical plants, coal co-firing plants, as well as steel and cement manufactures.

In response to Kyoto treaty obligations, governments throughout the world, including several states in the United States, have set up mechanisms to trade monetary carbon credits for human derived greenhouse gases. Of particular importance are CO_2 and CH_4 (methane) emissions. Monetary instruments such as the Emission Reduction Units (ERU) in the European Union allow industries to offset their CO_2 and CH_4 emissions by buying and selling carbon credit certificates. ASTM D6866 specification provides an accurate and precise monitoring and verification tool for carbon credit and carbon offsetting strategies.

b. ASTM D6522 - 00(2005): ASTM D6522 - 00(2005) is standard test method for determination of nitrogen oxides, carbon monoxide, and oxygen concentrations in emissions from natural gas-fired reciprocating engines, combustion turbines, boilers, and process heaters using portable analyzers.

c. ISO 14064 is the international standard: The ISO 14064 standards (published in 2006 and early 2007) are the most recent additions to the ISO14000 series of International Standards for environmental management.The ISO14064 standards provide governments, businesses, regions and other organisations with an integrated set of tools for programs aimed at measuring, quantifying and reducing greenhouse gas emissions. These standards allow organisations take part in emissions trading (chapter 18) schemes using a globally recognised standard. There are three parts in the standard as follows:

i. ISO 14064-1:2006 specifies principles and requirements at the organization level for *quantification and reporting of greenhouse gas (GHG) emissions and removals*. It includes requirements for the design, development, management, reporting and verification of an organization's GHG inventory.

ii. ISO 14064-2:2006 specifies principles and requirements and *provides guidance at the project level for quantification, monitoring and reporting of activities intended to cause greenhouse gas (GHG) emission reductions or removal enhancements.* It includes requirements for planning a GHG project, identifying and selecting GHG sources, sinks and reservoirs relevant to the project and baseline scenario, monitoring, quantifying, documenting and reporting GHG project performance and managing data quality.

iii. ISO 14064-3:2006 specifies principles and requirements and *provides guidance for those conducting or managing the validation and/or verification of greenhouse gas (GHG) assertions*. It can be applied to organizational or GHG project quantification, including GHG quantification, monitoring and reporting carried out in accordance with ISO 14064-1 or ISO 14064-2.

Carbon Dioxide Equivalent and Carbon Units

In Kyoto Protocol (3), emission figures of GHGs listed in Annex B conform to 'carbon dioxide equivalent emissions' of all GHGs listed in Annex I. Carbon dioxide equivalency is a quantity that describes, for a given mixture and amount of greenhouse gas, the amount of CO_2 that would have the same global warming potential (GWP), when measured over a specified timescale (generally, 100 years). The carbon dioxide equivalency for a gas is obtained by multiplying the mass and the GWP of the gas. For example, the GWP for methane is 21 and for nitrous oxide 310. This means that emissions of 1 million metric tonnes of methane a" 21 million tonnes of carbon dioxide and emissions of 1 million tonnes of nitrous oxide a" 310 million metric tonnes of carbon dioxide(chapter 4 & 5). Carbon dioxide equivalency (CO_2e) thus reflects the time-integrated radiative forcing.

The following units are commonly used for carbon equivalent:

- By the UN climate change panel IPCC: billion metric tonnes of CO_2 equivalent (GtCO_2eq).
- In industry: million metric tonnes of carbon dioxide equivalents (MMTCDE).
- For vehicles: g (gram) of carbon dioxide equivalents per km (gCDE/km).

ii. Conversion of Energy to CO_2-equivalent Emissions

A major percentage of the emissions is due to energy consumption Energy consumption is directly related to emission and can be converted to carbon equivalent emission units. Energy consumption figures are converted into emission units by using multiplying factor for the given application. The multiplying factors are provided by DOE in USA and similar agencies in other countries. In USA, carbon calculators use information from the US Department of Energy (Energy Information Agency) and other sources to develop an accurate assessment of carbon dioxide emissions emitted per energy type or use. EIA maintains an excellent website with easy to-understand and easy-to-access figures about all things energy related.[2]

The U.S. accounts for only 5% of the world's population, yet consumes 26% of the world's energy, making the country as the largest energy consumers in the world. Even though a large part of the energy is used for homes and businesses to power appliances and provide light, a majority of the energy is derived from burning of fossil fuels to run cars and industries. These fossil fuels produce

90% of the greenhouse gases that contribute to global warming, and the most excessive and harmful of these gases is carbon dioxide.

Every 24 hours Americans emit about 16 million tons of carbon dioxide into the air during the production of energy. This greatly increases the carbon footprints and its impact on the sustainability of the environment.Total US emissions were 7,122 million metric tons CO_2-equivalent in 2004. Dividing by the US population of 299 million, this equals 23.8 metric tons of CO_2-equivalent per person[1]. India's per capita GHG emissions are around a quarter of the global average(4.22 tons) at a value of 1.2 tons.

iii. Carbon Footprint

A *carbon footprint* is a "measure of the impact of human activities have on the environment in terms of the amount of greenhouse gases produced, measured in units of carbon dioxide". It is meant to be useful for individuals, nations and organizations to conceptualize their personal (or organizational) impact in contributing to global warming. A conceptual tool in response to carbon footprints are *carbon offsets*, or the mitigation of carbon emissions through the development of alternative projects such as solar or wind energy or reforestation. The carbon footprint is a subset of the *ecological footprint*, which includes all human demands on the biosphere. The concept is used to optimize the environmental performance of a single product (ecodesign) or to optimize environmental performance of a company, with final goal to becoming carbon neutral.The carbon foot prints are determined by using dedicated software packages,The present move within EU and globally to legislate for compulsory carbon emission reduction will make business to take effective steps to reduce the environmental impacts of their operations. This is the crucial step forward towards implementing cost effective reduction in greenhouse gases.

TWI[4] has developed a protocol based on best practice for a product level (or bottom up), which would assess the within the plant emission value for fabrication and assembly, with emission calculation approach validated by a range of established consultancies.The carbon footprint can be efficiently and effectively reduced by applying the following steps:

- *Life Cycle Assessment (LCA)* is also known as life cycle analysis, ecobalance, and cradle-to-grave analysis. LCA is the investigation and valuation of the environmental impacts of a given product or service caused or necessitated by its existence.
- Identification of hot-spots in terms of energy consumption and associated CO_2-emissions
- Where possible, changing to buying electricity from renewable sources.
- Optimize energy efficiency and, thus, reduction of GHG emissions.
- Identify solutions to neutralize the CO_2 emissions that cannot be eliminated by energy saving measures. This last step includes carbon offsetting; investment in projects that aim at the reducing CO_2 emissions, for instance tree planting.

As demand for energy grows over time, the total emissions must still stay within the cap, but it allows industry some flexibility and predictability in its planning to accommodate the extra requirement.

iv. Carbon Offset Management

Carbon offsetting is the act of mitigating or reducing ("offsetting") GHG emissions that cannot be directly reduced, by indirect means such as the planting of trees, investing in businesses engaged in the development of renewable alternatives such as wind and solar power, and supporting energy conservation initiatives, methane capture projects, and so on.

The idea of paying for emission-reductions elsewhere instead of reducing by own actions is also known from the closely related concept of emissions trading. However, in contrast to emissions trading, which is regulated by a strict formal and legal framework, carbon offsets generally refer to voluntary acts by individuals or companies that are commonly arranged by commercial or not-for-profit carbon-offset providers.

A wide variety of offset methods are in use while tree planting has initially been a mainstay of carbon offsetting, renewable energy and energy conservation offsets have now become increasingly popular. Airlines like North West (see chapter 12) have taken resort to tree planting as a part of carbon offset programme.

Carbon offsetting as part of a "carbon neutral" lifestyle has gained some appeal and momentum mainly among consumers in western countries who have become aware and concerned about the potentially negative effects of energy-demanding lifestyles and economies on the environment. This has contributed to the increasing popularity of voluntary offsets among private individuals and also companies. Offsets may be cheaper or more convenient alternatives to reducing one's own fossil-fuel consumption.

Carbon neutral refers to balancing the amount of carbon dioxide one releases with that amount that one can offset. To become carbon neutral any carbon dioxide that has been released through energy consumption or car emissions need to be compensated by using renewable energy sources or through buying of carbon credits. Kyoto Protocol has provision to buy any of the four different certified carbon units, viz., CER, ERU, RMU, and VER for becoming carbon neutral.

Emission Units

More than actual emissions units can be traded and sold under the Kyoto Protocol's emissions trading scheme. Each unit is equal to one tonne of CO_2. The transferable units belong to following types:

- Certified Emission Reduction (CER) a certified emission reduction (CER) generated from a *clean devolopment mechanism* project activity.
- Emission Reduction Unit (ERU) An emission reduction unit (ERU) generated by a joint implementation project.
- A removal unit (RMU) on the basis of land use, *land-use change and forestry (LULUCF)* activities such as reforestation.

- Verified Emission Reduction (VER).
- EU has a unified certified emission management and reduction scheme, known as, Certified Emissions Measurement and Reduction Scheme (CEMERS).

CEMARS and Carbon Neutral

The present move by EU and globally to legislate for compulsory carbon emission reduction will make the business and organisations conform to mandatory emission cut standards. CEMARS provides a highly credible solution to carbon reduction programme, which allows the organizations to measure, manage and report their organizational footprint via CEMARS (Certified Emissions Management and Reduction Scheme).

This is the crucial step forward towards implementing cost effective reduction in greenhouse gases. It enables businesses and organisations to measure their greenhouse gas (GHG) emissions in compliance with ISO 14064-1, understand their carbon liabilities, and put in place management plans to reduce emissions in their organisation and more widely through their supply chain. CEMARS has been developed for businesses and organisations, esp., for the large emitters and for whom offsetting is not a viable business proposition or have not yet made the business case for carbon neutral status

Emission limits-Kyoto Protocol

The concept of carbon credits came into existence as a result of increasing awareness of the need for controlling emissions. The IPCC has observed that:

Policies that provide a real or implicit price of carbon could create incentives for producers and consumers to significantly invest in low-GHG products, technologies and processes. Such policies could include economic instruments, government funding and regulation,

The emission allowance system functions as follows:

a. Kyoto Protocol, Annex I, contains a list of the countries alongside with agreed 'caps' or quotas on the maximum allowed amount of GHG emissions.

b. In turn these countries set quotas for the emission allowances from installations run by local business and other organizations, generically termed 'operators'. The main operators/sectors responsible for greenhouse gas emissions, are listed in Annex A, of Kyoto Protocol.

c. Countries manage this through their own national 'registries', which are required to be validated and monitored for compliance by the UNFCCC. UNFCCC or FCCC, (United Nations Framework Convention on Climate Change) is an international environmental treaty signed at the Earth Summit at Rio in 1992, latter ratified in the Kyoto Protocol. An important achievement of the treaty was to establish a *national greenhouse gas inventory* as count of GHGs emissions and removals, which must be regularly submitted by signatories to UNFCCC.

d. Each operator has an allowance of *credits*, where *each unit gives the owner the right to emit one metric tonne of carbon dioxide or other equivalent GHGs*.

e. Operators that have not used up their quotas can sell their unused allowances as *carbon credits*, while businesses that are about to exceed their quotas can buy the extra allowances as credits, privately or from the open market.

By allowing credits to be bought and sold, an operator can seek out the most cost-effective way of reducing its emissions, either by investing in 'cleaner' machinery and practices or by purchasing emissions from another operator who already has excess 'capacity'.

Kyoto Flexible Mechanisms –Carbon Projects

The organizations within the capped regime, has three choices for complying if they exceed their cap.

i. they can pay an alternative compliance measure or "carbon tax", a default payment set by the regulatory body. This choice is usually the least attractive given the ability to comply by trading.

ii. they can purchase *carbon credits* within an *emission trading scheme*. The trade provides an economic disincentive to the polluter, while providing an incentive to the less polluting organisation.

iii. final option is to invest in a *carbon project.* The carbon project will result in a greenhouse gas emission reduction which can be used to offset the excess emissions generated by the polluter.

A *carbon project* refers to a business initiative on a project that can receive funding because of the reduction of the emission of GHGs that will result. To prove that the project will result in real, permanent, verifiable reductions in GHGs, the proof must be provided in the form of a project design document and activity reports validated by an approved third party in the case of Clean Development Mechanism (CDM) or Joint Implementation (JI) projects.

However, the business and industry circles feel insecure due to following facts:

1. *Kyoto trading period* only applies for five years *between 2008 and 2012.* therefore there is general uncertainty as to the post-Kyoto negotiations on GHGs emissions, and thus adds to the risk and uncertainty to long term business planning.
2. Non-signatories to Kyoto protocols, such as, USA, Australia, China and India, who are responsible for a major proportion of GHG emissions, have no mandatory caps. Industries and business in capped countries paying for the carbon costs feel themselves at competitive disadvantage to the uncapped counterparts.

Guidelines for Emission Intensity

Emission intensity is the average emission rate of a given pollutant from a given source relative to the intensity of a specific activity; for example grams of CO_2 released per mega joule of energy produced, or the ratio of greenhouse gas emissions produced to GDP. Emission intensities are used to derive estimates of greenhouse gas emissions based on the amount of fuel burnt, or other practices. The related terms emission factor and carbon intensity are often used interchangeably, but "factors" exclude aggregate activities such as GDP, and "carbon" excludes other pollutants. One of the most important uses of emission factors is for the reporting of national greenhouse gas inventories under the United Nation Frame work Convention on Climate Change (UNFCCC). The Annex I Parties (Kyoto Protocol) are to report to the UNFCCC annually on their national total emissions of greenhouse gases in a formalized reporting format, defining the source categories and fuels.

UNFCCC has accepted the Revised 1996 IPCC Guidelines for National Greenhouse Gas Inventories, developed and published by the by IPCC as the emission estimation methods that must be used by the parties to the convention to ensure transparency, completeness, consistency, comparability and accuracy of the national greenhouse gas inventories. These IPCC Guidelines are the primary source for default emission factors. Recently IPCC has published the 2006 IPCC. The *2006 IPCC Guidelines for National Greenhouse Gas Inventories* are in 5 volumes, viz., vol.1 (General Guidance and Reporting), vol. 2 (Energy),vol.3 (Industrial Processes and Product Use),vol. 4 (Agriculture, Forestry and Other Land Use) and vol. 5 Waste). Guidelines relevant to all inventory guidelines volumes:

- Estimate carbon emissions in terms of the species which are emitted, including non-CO_2 species which eventually oxidizes to CO_2 in the atmosphere. If the emissions of these non-CO_2 gases contain very small amounts of carbon compared to CO_2 estimate and it may be more accurate to base the CO_2 estimate on the total carbon and not on individual gases.
- *Treatment of nitrogen (N) deposition*: The *GPG 2000* lists sources of anthropogenic nitrogen deposition that subsequently give rise to anthropogenic emissions of nitrous oxide (N_2O), but provides estimation methods only for a subset of these, associated with agricultural sources of ammonia (NH_3) and nitrogen oxides (NOx). The *2006 IPCC Guidelines are* extended to include all significant sources of

N deposition, including agriculture, industrial & combustion sources, with the ultimate N_2O emission attributed to the country responsible for the nitrogen originally emitted.

- *Relationship to entity- or project level estimates*: The Guidelines are intended to help prepare national inventories of emissions by sources and removals by sinks. Nonetheless, the Guidelines can also be relevant for estimating actual emissions or removals at the entity or project level.

According to the IPCC, if an activity is a major source of emissions for a country ('key source'), it is 'good practice' to develop a country-specific emission factor for that activity

Emission Intensity as a Measure of Country's Emissivity

The carbon intensity of the economy can largely be decomposed into two basic elements:[1] energy intensity, defined as the amount of energy consumed per dollar of economic activity; and[2] carbon intensity of energy supply, defined as the amount of carbon emitted per unit of energy. As illustrated by the formulas below, the multiplication of the two elements produces a numerical value for U.S. carbon intensity, defined as the amount of carbon dioxide emitted per dollar of economic activity:

Energy Intensity x Carbon Intensity of Energy Supply = Carbon Intensity of the Economy, or, (Energy / GDP) × (Carbon Emissions / Energy) = (Carbon Emissions / GDP).

From 2002 to 2003, the greenhouse gas intensity of the U.S. economy fell from 684 to 668 metric tons per million 2000 dollars of GDP (2.3 percent), continuing a trend of decreases in both carbon intensity and total greenhouse gas intensity. The carbon intensity and greenhouse gas intensity of the U.S. economy move in lockstep, because CO_2 emissions make up most of the total for U.S. greenhouse gas emissions. Energy-related carbon dioxide emissions represent approximately 83 percent of total U.S. greenhouse gas emissions.In 2008, the carbon intensity (tons of CO_2e per USD 1000) of United States and Russia was 21.5 and 15.9 compared to the values of 5.7 and 1.9 for China and India respectively. In EU, total CO_2 emissions from energy use were 5 % below their 1990 level in 2007[5]. Over the period 1990-2007, CO2 emissions from energy use have decreased on average by 0.3 %/year although the economic activity (GDP) increased by 2.3 %/year.

The carbon management has been a standard practice in the mandatory regimes and a part of CSR (see chapter 19) for others. Management guidelines for low carbon business are provided in various books and literatures[5].

Carbon Thinking

'Carbon thinking' is required by corporations in all spheres of activities to make the organisation as a whole carbon neutral. Carbon thinking must permeate through full range of company activities from operations and trading to investment planning; and from legal and environmental compliance to tax and accounting.

Green procurement policy ensures that materials obtained are reasonably carbon neutral. Energy efficient operations need less energy per unit produced. The objective should be to produce more with less energy consumption or emission.

Mandatory carbon reductions by corporations belonging to signatories to Kyoto Protocol need to comply with the imposed limits. Legal aspects of violating some of the regulatory measures to save environment need to be evaluated in terms of the activities of the corporation. Accounting should take note of the financial results due to carbon management practices followed by the organization.

Corporations will need to stress test on green aspects in their longer-term planning against a range of possible regulatory and market scenarios. For example, automobile manufacturers in EU countries need to go for more fuel efficient cars in the future because of newer regulations on the anvil. Future markets would be for the hybrid cars or cars based on fully clean fuel.

Companies will need to think about ways to communicate carbon risks and opportunities to investors, customers and financial institutions. Transparency in the carbon risks and opportunities would save the company from getting future disaster.

Companies and institutions can benefit from the development of an emission accounting, measurement, and tracking process. Such a system requires a detailed energy audit and metering study, which provides the basis for emission accounting and can employ one of several generic protocols now available. In 2005, EU legislation aimed at capping carbon emissions generated billions of euros in assets and liabilities. For the companies affected, a choice must be made to cut emissions and /or save energy, invest in international emission reduction projects, or trade emissions. With this choice come new risks and opportunities. The topic would be discussed in details in chapter 19 on CSR.

REFERENCES

1. Source: *Energy Information Agency*, US Census Bureau.
2. US: EIA, *Department of Energy*, www,eia.doe.gov
3. Kyoto Protocol—*The United Nations Framework Convention on Climate Change.* United Nations unfccc.int/resource/docs/convkp/kpeng.pdf
4. Carbon Footprints—Product life cycle/ Carbon Foot priting,TWI(The Welding Institute,UK) Connect magazine, issue 153,March/April,2008
5. Energy Efficiency Indicator in Europe,http://www.odyssee-indicators.org/
6. The New Business Climate –A Guide to Low Carbon Business and Business Performance; a book published by Rocky Mountain Institute. climateregistry.org/resources/docs/.../RMIKEMA_CAR_1105.pdf

12

Carbon Sequestration and Geo-engineering

Carbon sequestration is the process of removing carbon from the atmosphere and depositing it in a reservoir. When carried out deliberately, this may also be referred to as carbon dioxide removals, which is a form of geo-engineering[1]. Carbon dioxide is naturally captured from the atmosphere through biological, chemical or physical processes. The natural processes include the capture of atmospheric carbon dioxide in a solid material (such as growing trees, other vegetation, and soils) or a carbon sink through biological or physical processes, such as photosynthesis. The natural carbon sinks for carbon sequestration are discussed in chapter 6 & chapter 15. In this chapter, the anthropogenic sequestration techniques are to be discussed. Some of the anthropogenic techniques exploit the natural processes, while some use entirely artificial processes[1].

Geo-engineering can be defined as the human efforts to replicate natural processes in order to counteract the effects of global warming. Geo-engineering techniques include direct methods of reducing carbon dioxide by carbon sequestration[1] or indirect methods, such as, ocean iron fertilization to produce phytoplankton which would act as carbon sink. Alternatively, proposed geo-engineering techniques include solar radiation management by sulphur aerosols in stratosphere and cloud reflectivity enhancement[2]. Most techniques have at least some side effects. In this chapter geosequestration, the most promising method is to be discussed in details. Ocean iron fertilization shall be discussed in Chapter 15.

Geosequestration

The fossil fuels, such as, coal, oil and natural gas, currently supply around 85% of the world's energy needs. The International Energy Agency predicts that fossil fuels will continue to be heavily used for many years to come. The burning of fossil fuels is a major source of excess CO_2, the gas that has most contributed to the increased concentration of greenhouse gases in the atmosphere. The *geosequestration* involves the capture and long-term underground storage of carbon dioxide[1,5]. Carbon dioxide produced by coal-fired power generation plants and other industrial sources is compressed to form a liquid and injected into deep underground geological formations, such as, saline aquifers, coal seams, and used oil and gas reservoirs. The geosequestration enables the long-term geological storage of CO_2. Other techniques of carbon storage include, new technology to capture carbon in aluminum industry[3], carbon sequestration by algae[4], and carbon dioxide absorption by new metal-organic crystals[6].

Carbon Capture & Storage (CCS) Technology

The coal based power generating plants with long life (50 years and beyond) and their growing numbers are becoming major source of carbon dioxide emission. To prevent the carbon dioxide to escape in the atmosphere, the technology to 'capture' carbon and 'storing' elsewhere, has been developed.

The stages in the process of geosequestration of carbon dioxide from the combustion products are as follows:

i. *Enrichment of carbon dioxide* in the flue gas
ii. *Separation or capture of carbon dioxide* from the flue gas.
iii. *Compression of carbon dioxide* to a lower volume for ease of transportation and storage.
iv. *Storage of compressed carbon dioxide* in leak proof areas.

Carbon dioxide Enrichment of Flue Gas in Power Plants

The main source for bulk capture of CO_2 is electricity generation from fossil fuel sources. Industrial processes such as, natural-gas processing, ammonia production, and cement manufacture, produce relatively small quantities of CO_2 and thus are not suitable for bulk storage. A far larger source of CO_2, accounting for one-third of total CO_2 emissions in Australia, is fossil-fuelled electricity generation.

The capture of CO_2 from a stationary source, such as power plant, involves separating and purifying CO_2 from the bulk of the flue gas stream before geological storage. Due to the presence of large quantities of nitrogen, the partial pressure of the carbon dioxide in the flue gas is very low in the fossil fuel based power plants. The direct recovery of the carbon dioxide from such flue gas, known as "flue gas approach", is not a techno-economically viable proposition. Therefore, alternate strategies involving enrichment of carbon dioxide in the flue gases are used in capturing CO_2 from electricity generation plants. The processes for CO_2 enrichment in flue gas include the followings:

A. Post-combustion, using air for combustion: Pulverized coal using the flue gas approach presents the largest economic hurdle to CO_2 sequestration. The flue gas approaches in use today require clean-up of the NOx and SO_2 prior to CO_2 separation. If the sinks are tolerant to NOx and SO_2, it is possible to eliminate separate control steps and sequester the NOx and SO_2 along with the CO_2, resulting in a zero emissions power plant.

B. The "hydrogen" or "syngas" approach or pre-combustion or hydrogen route, generally referred to as Integrated Gasification Combined Cycle or IGCC, Integrated coal gasification combined cycle (IGCC) plant is an example of the hydrogen route. Coal is gasified to form synthetic gas (syngas) of CO and H_2. The gas then undergoes the water-gas shift, where the CO is reacted with steam to form CO_2 and H_2. The CO_2 is then removed along with the hydrogen for sending to a gas turbine combined cycle. A similar process is available for natural gas, where the syngas is formed by steam reforming of methane.The hydrogen route opens up opportunities for "polygeneration", where besides electricity and CO_2, additional products are produced. For example, instead of sending hydrogen to a turbine, it can be used to fuel a "hydrogen economy". In addition, syngas is an excellent feedstock for many chemical processes.

C. Pre-combustion, oxyfuels, the "oxygen" approach: The major component of flue gas is nitrogen, present as 75% of air feed, without which CO_2 capture from flue gas would be greatly simplified. Therefore use of oxygen instead of air for combustion would lead to easier separation of CO_2 from flue gas. However, combustion with oxygen produces too high a temperature for today's materials, so some flue gas must be recycled to moderate the temperature. Applying this process is easier for steam turbine plants than gas turbine plants. In the former, relatively straightforward boiler modifications are required, for the latter, much more.

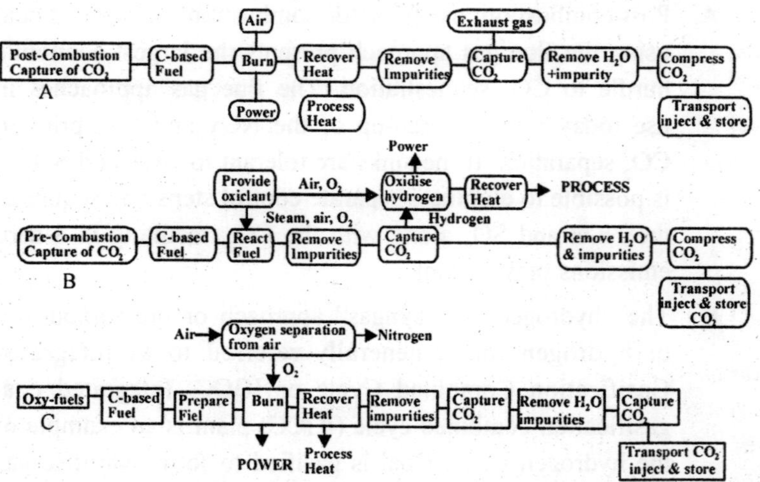

Fig. 11.1 Flowsheets of Carbon dioxide Capture & Storage Processes for Power Plants using Fossil Fuels; A & B = Post- & Per- Combustion; C= Oxy-fuel Combustion

Separation and Capturing of Carbon dioxide

The separation and capture of carbon dioxide operations take place at the power plant site, including compression. After separation, CO_2 is generally compressed to the order of 100 atm, for easy tranportation.

The idea of separating and capturing CO_2 from the flue gas of power plants did not start with concern about the greenhouse effect. Rather, it gained attention as a possible economic source of CO_2, especially for use in *enhanced oil recovery (EOR)* operations where CO_2 is injected into oil reservoirs to increase the mobility of the oil and, therefore, the productivity of the reservoir. Several commercial CO_2 capture plants were constructed in the late 1970s and early 1980s in the US. Also, this process is used to produce *CO2 for carbonation of brine.* Several more CO_2 capture plants were subsequently built to produce CO_2 for commercial applications. Some of these plants took advantage of the economic incentives in the Public Utility Regulatory Policies Act (PURPA) of 1978 for "qualifying facilities".

Four main techniques for separation of carbon dioxide from flue gas are as follows:

a. *Absorption*, where CO_2 is selectively absorbed into liquid solvents

b. *Membranes*, where CO_2 is separated by semi permeable plastic or ceramic membranes

c. *Adsorption*, where CO_2 is separated using specially designed solid particles and

d. *Low Temperature Processes*, where separation is achieved by chilling and/or freezing the gas stream.

Plants based on membrane separation, cryogenic fractionation, and adsorptions using molecular sieves are less energy efficient and more expensive than chemical absorption. The primary difference in capturing CO_2 for commercial markets versus capturing CO_2 for sequestration is the role of energy. In the former case, price is sole consideration not the quantum of energy used. In the latter case, minimum quantity of energy is to be used to avoid causing more CO_2 emissions. Therefore, capturing CO_2 for purposes of sequestration requires more emphasis on reducing energy inputs than the traditional commercial process. Virtually all commercial processes for CO_2 separation for sequestration are based on absorption in liquid solvents. The solvents used may be categorized into two types:

a. *chemical solvents*, such as aqueous solutions of monoethanolamine (MEA) or potassium carbonate, where the mechanism of absorption is via a reversible chemical reaction. The absorption rate of CO_2 in unpromoted hot potassium carbonate solutions is vastly improved in solutions promoted with diethanolamine (DEA) or with sterically hindered amines.

b. *physical solvents*, such as, methanol used in Rectisol® or dimethyl ethers of polyethylene glycols used in Selexoll®, where the absorption of gases occurs without chemical reactions.

The hydrogen route for carbon enrichment allows for a CO_2 removal process by a physical solvent process like Selexsol, which is much less energy intensive than the MEA process, because capture takes place from the high pressure syngas as opposed to the atmospheric pressure flue gas.

MEA Method

The most widely used method till date for capturing carbon dioxide is based on chemical absorption with a monoethanolamine (MEA) solvent. MEA was developed over 60 years ago as a general, non-selective solvent to remove acid gases, such as CO_2 and H_2S, from natural gas streams. The process allows flue gas to contact an MEA solution in the absorber. The MEA selectively absorbs the CO_2 and is then sent to a stripper. In the stripper, the CO_2-rich MEA solution is heated to release almost pure CO_2. The lean MEA solution is then recycled to the absorber. The technique has been in use for more than four decades in producing endothermic protective atmospheric gas (4%CO + N_2), by burning ATS grade, low sulfur kerosene. The protective endothermic gas is used in continuous annealing furnace for bright annealing of cold rolled steel strips.

Sequestration of Carbon dioxide

While the capture of CO_2 for geosequestration is a relatively new concept, CO_2 capture for commercial markets has been practiced in Australia and overseas for many years. CO_2 is captured from natural gas wells in south-east South Australia, near Mt Gambier and in southern Victoria, near Port Campbell. The CO_2 is then used for various commercial processes including carbonation of beverages and dry-ice production.

After capturing, CO_2 can be stored in leak proof areas for a long period, such as, oil and gas fields, which have preserved oil and gas for million years.Also can be used as excess gas for producing oil by forcing from ground by injected CO_2. British Geological survey says there is an estimated 6.2-8.5 Gt of storage capacity in the UK's on- and off shore oil and gas fields-which could hold in entire CO_2 emissions for 10-15 years. UK is expected to build first CCS demonstrator plant by 2014. In USA the first coal based power project to capture carbon dioxide (pumping to underground storage), based on CCS technology, due to come up in Illinois, has been shelved till 2015[5].

In the United States, CO_2 capture at power plants using chemical absorption solvent has been practiced since the late 1970s, with the captured CO_2 being used for enhanced oil recovery.There are plans in the United States to build the world's first integrated gasification combined cycle plant, known as FutureGen, that will not only produce electricity but also hydrogen fuel, with the CO_2 generated in the process being captured and sequestered.

Following capture, CO_2 is usually transported from the source, such as a power station, to the geological storage site in a compressed form via a pipeline. It can also be transported by truck, rail or ships depending on the location of both the source and the geological storage site and injected via pipeline deep underground.

Comparing Offshore Geological Storage of CO_2 to Ocean Storage of CO_2

In both offshore geological and ocean storage of CO_2, include, capturing the gas from a stationary emissions source such as a power plant or other industrial facility and then transporting the highly compressed CO_2 offshore via a sub-sea pipeline or ocean tanker. There is, however, a major difference between offshore geological sequestration and ocean sequestration in the way in which the CO_2 is stored.

Offshore geological storage involves the CO_2 being injected into a geological formation deep beneath the sea bed where it will be stored for thousands of years, isolated from the ocean water. In the case of ocean storage, the CO_2 is injected directly into the water column either at mid-depth (1500 to 3000 metres), where it dissolves in the ocean waters, or at greater depths (below 3000 metres), where it forms a deep CO_2 lake. Offshore geological storage has been successfully demonstrated at Statoil's Sleipner field in the North Sea (about 250 km off the coast of Norway). At Sleipner, CO_2 is separated from produced natural gas and stored in a deep saline formation about 1000 metres beneath the seabed. No ocean sequestration demonstration projects exist so far.

New Technology to Capture Carbon in Aluminum Industry

In 2007, ALCOA, multinational aluminum major, launched a new 'carbon capture' technology to its Kwinana aluminum refinery in Western Australia. The process for capturing includes mixing of carbon dioxide with bauxite residue, a byproduct of aluminum metal extraction process.This locks up large amount of the greenhouse gas generated by the aluminium extraction plant, which would otherwise be released to atmosphere. The Kwinana carbonation plant will lock up 70,000 tons of CO_2 a year, the equivalent of eliminating the emissions of 17,500 automobiles. The new technology has the potential to deliver significant global greenhouse benefits and will contribute to a reduction in the aluminum industry's environmental footprint. By mixing carbon dioxide with bauxite residue,the compound formed has similar alkalinity level as found in alkaline soil. The new mixture can be used as additive to improve soil, for road foundation and building material.In addition to its greenhouse benefits, ALCOA's residue carbon capture process delivers other economic, environmental and social benefits. A major benefit incldes biological removal of sodium oxalate through biodegradationn by natural bacterial activities in the carbonate residue. Normal process for removal is heating in kiln, an energy consuming thermal process.

Carbon Sequestration by Algae

Engineers have designed a simple, sustainable and natural carbon sequestration solution using algae. A team at Ohio University created a photo bioreactor that uses photosynthesis to grow algae, passing carbon dioxide over large membranes, placed vertically to save space. The carbon dioxide produced by the algae is harvested by dissolving into the surrounding water. The algae can be harvested and made into biodiesel fuel and feed for animals. A reactor with 1.25 million square meters of algae screens could be up and running by 2010[4].

Carbon Dioxide Absorption by new Metal-organic Crystals[6]

Scientists at University of California have made metal-organic crystals, named as zeolite imidazolate, or ZIF, capable of soaking up carbon dioxide gas like a sponge. The crystals are non-toxic and would require little extra energy from a power plant, making them an ideal alternative to current methods of CO_2 filtering. The porous structures can be heated to high temperatures without decomposing and can be boiled in water or solvents for a week and remain stable, making them suitable for use in hot, energy-producing environments like power plants. The highly porous crystals have "extraordinary capacity for storing CO_2"; one litre of the crystals could store about 83 litres of CO_2. Estimates from United Nation's energy and climate experts have pegged the cost of capturing CO_2 between $25 US and $60 US a tonne for conventional coal-fired plants.

Clean coal

The coal industry uses the term "clean coal" to describe technologies designed to enhance both the efficiency and the environmental acceptability of coal extraction, preparation and use, with no specific quantitative limits on any emissions, particularly of carbon dioxide.

Solar radiation management

The techniques include, creating stratospheric *sulfur aerosols* to scatter or reflect solar radiation, using *pale-colored roofing and paving materials*, *cloud reflectivity enhancement* by using fine salt water spray to whiten clouds and increase cloud reflectivity, and *space sunshade* by obstructing solar radiation with space-based mirrors or other structures Limited trials have been conducted on cool *roof* and *stratoscopic sulfur aerosol spray.*

Regulation of geo-engineering

According to Ken Caldeira, a professor of climate science at Stanford University, reducing greenhouse gases will cost around 2 percent of the gross domestic product, while geo-engineering (by putting reflective aerosols into the upper atmosphere) will cost about one-thousandth of that. However, geo-engineering needs to be regulated in order to prevent adverse effects; that there should be independent assessment of the impacts of geo-engineering research proposals before actual use of the technologies. These are views expressed by the scientists at a recent conference at Asilomar, and also echoed by scientific communities, including Royal Society and the Academy of Sciences for the Developing World.[7] However, large-scale geo-engineering projects (excepting ALCOA process) are yet to be commercialized.

Summary

1. Geosequestration of emitted carbon dioxide includes capture and storing of the gas. Chemicals (MEA) reaction and physical (methanols) absorption methods are used for CO_2 capture. Recently developed media to capture includes, bauxite waste (ALCOA), algae, and zeolite are used to capture. CCS process needs underground storage of the gas. To increase the carbon concentration in the emission, either oxygen is used for combustion or pre-combustion

gasification is carried out with steam, thus producing water gas for subsequent combustion.

2. Out of proposed geosequestration projects, till date only project implemented is that of ALCOA.

REFERENCES

1. Carbon-Sequestration; Wikipedia;http://en.wikipedia.org/wiki/Carbon sequestration.
2. David Schnare, *Environment & Climate News*, Dec, 2007, The Heartland Institute.
3. *Capture carbon in aluminum industry*, Span, April-March, 2008, p. 17.
4. Carbon sequestration by algae, Science Daily, April 1, 2007.
5. Trouble in America's leading CCS project, *The Economist*, feb2, 2008.
6. Metal-organic frameworks with high capacity and selectivity for harmful gases. David Britt, David Tranchemontagne, & Omar M. Yaghi,Proc Natl Acad Science, USA, v105(33); Aug 19, 2008.
7. *The Economist*, Geoengineering. April 3, 2010, p. 73-74.

13

Fossils and Alternative Energy Sources

The basic science has taught us that *'the energy cannot be created nor destroyed but can be transformed from one type to another'*. For example, the stored energy in carbon can be released by burning fossil fuel, and the heat energy thus formed can be used to produce either steam or high energy gas, which in turn can be used to drive steam or gas turbines for power generation or to utilize as motive power for driving transport systems. Energy transformation from one form to another involves dissipation (loss) of substantial quantity of energy in some other form.The transformation or power generation efficiency is always below 100%.

The major source of GHGs till date is that produced by burning of fossil fuel to generate electricity and to run transports.Thus the thrust areas to be managed include, apart from development of alternatives, the improvement in power generation and utilization efficiency. Asia has the fastest projected regional growth in electric power generation worldwide, averaging 4.4 percent per year from 2006 to 2030, which is led by China and India[1]

Alternative sources of energy based on natural resources, such as wind, solar beam and water are free from GHG emissions. The reduced GHG fuels, such as biofuels produce less GHG than the normal fossil fuels like coal. Biofuels are made from renewable and sustainable sources, but not free from GHG emissions. However biofuel sources can be made carbon neutral by growing plants with capacities to absorb carbon dioxide of equivalent to that of emitted amount. Liquid biofuels are mostly used in transportation sector, which would be covered in chapter 14.

Excepting few, such as, hydroelectric, major alternatives are yet to enter in a big way in the commercial arena. However smaller alternatives, such as, wind, solar and biofuel collectively are growing at a fast rate, and expected to play major role in the energy generation. These renewable and sustainable alternatives should provide longtime energy security with no or negative effect on global warming.

Global Power Generation Scenario[1]

The US Energy Information Administration's figure of global energy consumption in 2004 is estimated to be 15TW (in SI units, T= tera= 10^{12} = trillion). (Tab.12.1)(ref.1).Fossil fuels supply 86% of world's energy. Major share belongs to oil (37.33%); followed by coal (25.33%) and then gas (23.33%).The proportion varies with the availability of resources. In China & India, coal has the major share in fossil fuel based power generation industries. Improvement in power generation efficiency is the major task involved in reducing GHG emissions in the fossil fuel based power generation industries.

Table 12.1 Global Energy Consumption

Fuel type	*Power in TW*	*Energy% Per year*
Oil	5.6	37.33
Gas	3.5	33.33
Coal	3.8	25.33
Hydroelectric	0.9	6
Nuclear	0.9	6
Geothermal, wind, solar, wood	0.13	0.86
Total	15	99

Amongst the alternative resources, hydro and nuclear accounts for 12% and rest 1% belong to wind, solar and biofuel.The energy conversion efficiency in alternate energy segment is equally important to cater to growing need to energy at a lower average emission figure.

Carbon dioxide Generation by Fossil fuel based Power Plants

In the last forty years, the use of fossil fuels has continued to grow and their share of the energy supply has increased. In the last couple of years, coal, which is one of the dirtiest sources of energy, has become the fastest growing fossil fuel. China plans to build or expand 199 coal-fired facilities in the next decade, compared with 83 in United States.[1] India plans to build 25 coal based mega-power plants of 4000MW each by 2012[2]. The International Energy Agency[1] predicts that fossil fuels will continue to be heavily used for many years to come. In 2006, coal's share of generation was an estimated 79 percent in China and 71 percent in India, which are likely to decline to 56 percent in India and 75 percent in China in 2030[1]. The achievement of the emission target would be possible with the use of energy efficient thermal plants and alternate energy sources in this fast growing sector.

The recent IEA report (1a) includes estimated *energy related carbon dioxide emissions in 2010 as record 30.6 Gigatonnes* (Gt), compared to CO_2 emission of 29.3 Gt in 2008, which was followed by a dip in 2009 due to the global recession. To achieve the goal of 2°C limitation, as agreed in Cancun in 2010, global energy-related emissions in 2020 must not be greater than 32 Gt. This means that over the next 10 years, emissions rise should be less in total than they did between 2009 and 2010. In terms of fuels, according to IEA, 44 percent of the estimated CO_2 emissions in 2010 came from coal, 36 percent from oil, and 20 percent from natural gas. There is thus a greater need to manage emissions in the energy front.

Power plants account for 40 percent of U.S. greenhouse gas emissions and 25 percent of the world. China, South Africa and India host the world's five dirtiest utility companies in terms of global warming pollution[3]. The world's top-ten power sector emitters in absolute terms are China(no.1), the United States, India (no.3),

Russia, Germany, Japan, the United Kingdom, Australia, South Africa, and South Korea (no.10). If the 27 member states of the European Union are counted as a single country, the E.U. would rank as the third biggest CO_2 polluter, after China and the United States.

Reduction of Carbon dioxide Emission in Fossil Fuel Plants

Coal-based power plants have the major share in producing electricity compared to any other single source. The life of a thermal power plant ranges between 50 to 100 years. Hence it is not possible to close down these plants just because they are the worst emitters of carbon dioxide. The improvement in the generating efficiency of coal-based power plants holds the key to minimizing emissions per unit of electricity generated. The strategy involved in this sector is to address the following issues:

i. **Improvement in energy generation efficiency by prolonging plant life:** The hard minerals (ash) content of coal (around 40% in Indian coal) cause damage to grinding equipments and boiler tubes by wear of surface materials. In Europe, coal contains high sulphur which leads to damage of the boiler tubes by corrosion[4]. The progressive wear and corrosion of the power plant components would result in malfunctioning of equipments leading to premature failure.As a result of poor plant performance,the energy generating efficiency can be substantially reduced. The direct and consequential losses in the production can be anywhere between 20% to 45% in a year. Tribo-science based wear prognosis and surface engineering[4,5] have in recent past, been able to provide solutions to minimize wear & corrosion and thus enabling improvement in the plant performance and energy generating efficiency. Plant life cycle analysis and extension through appropriate maintenance and repair technology are being increasingly used to further improve the plant efficiency[6].

ii. **Improvement of thermal energy generating efficiency: through innovative practices:** In a thermal power plant, heat or thermal energy produced by burning fossil fuel or nuclear fission is used to convert water into steam in a boiler. The steam is used to drive a turbine (in the generator) to generate electricity. In a thermal power plant out of 100 units of heat produced, typical heat loss figures include 10% in stack gases, 50% in cooling water, and 3% in transmission. Therefore the resultant efficiency of conversion is 37% maximum under normal conditions of operation. The average global efficiency of coal-fired plants is currently 27% compared to 45% for the most efficient plants. Improvements in the efficiency of *thermal power plant using coal* can be achieved with the adoption of following technological innovations[7]:

a. ***Fluidized Bed Combustion:*** In fluidised bed combustion, pulverized coal is burnt in a reactor comprised of a bed through which gas is fed to keep the fuel in a turbulent state. The fluidised state of the coal powder improves combustion efficiency, heat transfer and recovery of waste products. The higher heat exchanger efficiencies and better mixing in FBC system allows operation at lower temperatures than conventional pulverised coal combustion (PCC) systems.

b. ***Integrated Gasification Combined Cycle:*** An alternative to use of pulverized coal as 'fluid' in a fluidizd bed reactor is to convert coal into a fluid (gas) in a gassifier where carbon from coal reacts with oxygen and steam to produce the syngas, which is mainly H_2 and carbon monoxide (CO). IGCC plants use a gasifier to convert coal to syngas, which drives a combined cycle turbine. Waste heat from the gas turbine is recovered to create steam which drives a steam turbine, producing more electricity. Hence a combined cycle system produces more power with same quantum of coal.

c. ***Supercritical & Ultrasupercritical Boilers:*** The term 'supercritical' refers to thermodynamically equilibrium conditions where both liquid (water) and gaseous (steam) phases exist together as a homogenous fluid The supercritical stage in water exists at a pressure over 22.1 MPa (3207 psi) at the boiling point. Up to an operating pressure of around 19 MPa in the evaporator part of the boiler, the cycle is subcritical. This means, that there is a non-homogeneous mixture of water and steam in the evaporator part of the boiler. In this case, a drum-type boiler is used because the steam needs to be separated from water in the drum of the boiler before it is superheated and led into the turbine.

In the evaporator, at an operating pressure greater than 22.1 MPa, the cycle medium is a single-phase homogeneous fluid. Once-through boiler without drum seperator is therefore used in supercritical cycle. Since the once through boiler does not rely on the density difference between steam and water to provide proper circulation and cooling of the furnace enclosure tubes, it can be operated at pressures above the supercritical point.

Supercritical power generation units feature once-through boilers designed to operate with pressures from 3,500 to 4,000 psi, versus 1,800 to 2,500 psi for subcritical boilers. The temperature in subcritical, supercritical and ultra supercritical range are approximately at 538°C, 566°C & upto 700°C respectively[7a]. Ultra-supercritical power generation uses higher pressures in the supercritical range. *Higher firing temperatures and pressures improve conversion efficiency, defined as more electricity generated per BTU of coal consumed.* The significantly high enrgy conversion efficiency and results in similar reduction in fuel consumption, carbon dioxide, SO_2 and NOx (acid rain) emissions per megawatt of power output. The increased efficiencies translate into reduced fuel costs and emissions[8].

The average efficiency of a coal-fired power plant operating in subcritical range(19 MPa, 2757 psi) is around 27% with an emission figure of 1259gm CO_2 per kWh of electricity generated. In a supercritical operation (24 MPa, 3500 psi), the efficiency can be as high as 45% and the corresponding emission figure can be as low as 750gm CO_2 per kWh of electricity generated. In ultra-supercritical operation (27.5 MPa, 4000 psi) the efficiency can go up beyond 55% and carbon dioxide emission figures can be lower than 500g per kWh of electricity generated. The improvement in percent efficiency in supercritical from subcritical is 18% with reduction in emission figure by 40.4%. Similarly from super to ultrasuper (27. 5 MPa, 4000 psi) efficiency improves by 10% and corresponding emission reduction by 33.33%. One percentage point improvement in the efficiency of a conventional pulverized coal combustion plant results in a 2-3% reduction in CO_2 emissions[7,7a].

A series of fossil fuel based, high efficiency Ultra-Mega-power generating units has been a part of the ambitious project, to narrow the gap between supply and demand in India, a country of chronic power deficit. This would entail a creation of an additional capacity of at least 100,000 MW by 2012. The UM plant is considered as belonging to 'Clean Development Mechanism' by the organization that administers the Kyoto Protocol. This allows industrialized nations to invest in the plant as an alternative to domestic emissions reductions(2).The first UM-plant, the Tata Ultra Mega, a $4.2 billion power plant is being built near at Mundra, Gujrat, India

iii. Use of Clean Coal Technology & Capping of Carbon dioxide (see chapter11)

Renewable Sources for Power Generation

The renewable resources/sources for generating electricity can be divided in two-groups, as follows:

i. *No-GHG sustainable natural energy sources:* The energy sources belonging to this group are omni-present in nature and replenished by nature as soon as depleted, hence also called *perpetual resources.* Major alternate sources, such as, water (hydel power), sunlight (solar), geothermal and wind (wind power) do not, in real sense, get depleted, when used for power generation. Also one need not pay for getting access to and utilizing these resources, unlike fossil fuels or even alternative like bio-mass. These sources do not diectly emit GHGs while used for power generation. However even in this non-GHG emitting segment, the improvement in *energy conversion efficiency* is important to make the process techno-economically viable, to derive maximum benefit by producing more of clean energy utilizing less resources and to have minimum effect on the ecology per unit of power produced.

ii. *Reduced- GHG sustainable alternate fuels:* Biomass producing biofuels belongs to this group.GHG emissions on burning biofuel are less than that of fossil fuels. However, the biomass energy require wise management if they are to be used in a sustainable manner. Biofuels based on plantation can be possibly made carbon neutral, only if, carbon dioxide absorbed by plants in their growth process becomes equal to carbon dioxide emitted by biofuel made from same quantity of plants.

Renewable energies are also sustainable energy as they generally contribute to world *energy security*, reducing dependency on fossil fuel resources, and providing opportunities for reducing GHGs.The International Energy Agency has defined three generations of renewable energy technologies, as follows:

First-generation technologies resulting from the industrial revolutions at the end of the 19th century include hydropower, biomass combustion, and geothermal power & heat. Some of these technologies are still in widespread use.

Second-generation technologies include solar heating & cooling, wind power, modern forms of bioenergy and solar photovoltaic. Since 1980s, these are gradually being introduced in the commerial markets.

Third-generation technologies include advanced biomass gasification, biorefinery technologies, concentrating solar thermal power, hot-dry-rock geothermal power, and ocean energy. Some of them, like solar thermal have grown rapidly from the beginning of new century.

i. Water power- Hydropower

Hydroelectric power, which underwent extensive development during growth of electrification in the 19th and 20th centuries, is experiencing resurgence of development in the 21st century. The areas of greatest hydroelectric growth are the booming economies of Asia. China is the development leader; however, other Asian nations are installing hydropower at a rapid pace. This growth is driven by much increased energy costs – especially while using imported raw materials (coal, oil) — and widespread desires for more domestically-produced, clean, renewable, and economical generation.

Energy in water in the form of motive energy can be used for power generation by running a turbine. Since water is about 800 times denser than air, even a slow flowing stream of water, or moderate sea sweil, can yield considerable amounts of energy. The total world production of hydroelectric powder in 2005 is 2994 TWh. China, Canada & Brazil are top three amongst the top ten producers. The production in India is around 25% of that of China (9). However Norway leads the world with clean hydro-electricity constituting virtually the total domestic power production (98.9%).Although rich in fossil fuel resources, Brazil and Venezuela have made significant contribution in using clean energy by having share of hydro in domestic power generation as 83.7% and 73.9% respectively. Canada (57.9%) and Sweden (46.9%) have around half of their domestic production in hydro.

Table 12.2 Shares of Top Ten Hydro-electricity Producers

Country	*Generation (TWh)*	*% of World Total*
China	397	13.3
Canada	364	12.1
Brazil	337	11.1
USA	290	9.3
Russia	175	5.8
Norway	137	4.6
India	100	3.3
Japan	86	2.9
Venezuela	75	2.5
Sweden	73	2.4
Rest of the World	960	32.1

Ref: International Energy Agency, Key World Statistics, 2007

All other coutries have less than 19% share of domestic generation. In India, hydroelectric has a share of around 14% of total power generating capacity (tab.12.3)[9] Northern states in India and the adjoining countries like Nepal and Bhutan are major producers on hydroelectric power in the sub-continent.

Table 12.3 Share of hydro in the Country's total power generation

Country	*Hydro % of total domestic generation*
Norway	98.9
Brazil	83.7
Venezuela	73.9
Canada	57.9
Sweden	46.0
Russia	18.3
China	15.9
India	14.3
Japan	7.8
USA	6.8
Rest of the world	13.9
World average	16.4

Ref: international Energy Agency, Key World Statistics, 2007

High Silt Erosion of Hydroturbines in Himalyan Regions- Nature's Revenge

The extensive deforestation in the northern Himalayan region of India and Bhutan, has resulted in severe soil erosion by rain water and falls. (fig12.1).The silts (mainly anthropogenic) formed by erosion of deforestated soil, are carried by flowing water jets or columns to join the downstream river. The hydroelectric power plants on these rivers are subjected to severe erosion by high concentration of silts (with high proportion of quartz) in the water. The runner blades and guide vanes are subjected to heavy damage, leading to substantial reduction in power generating capacities. The silt erosion damage scar on worn turbine in fig12.2 is similar to that of deforested soil erosion scar fig.12.1 (ref.5).

In the hydroelectric power plants in the Himalyan regions in India and Bhutan, the erosion can be so severe as to remove half of the turbine's runner blades in six months' time.With the decreasing size of the runner blades there is a proportionate drop in power generation. The downtime for repair and re-blading is around three months in a year,the period in which there is no power generation. The total loss in generating capacity can be as high as 40% per year in the worst affected plants.

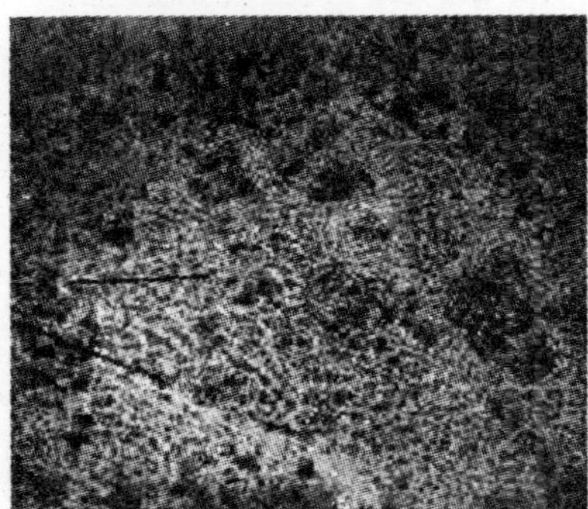

Fig. 12.1 Deforested Hills, soil Erosion, Chuka Bhutan (photo by authors, 4 June, 2008)

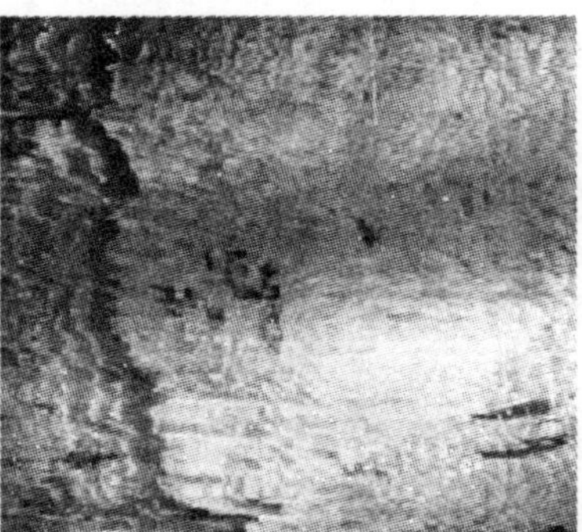

Fig. 12.2 Eroded Turbine, Hydroelectric Project Chuka, Bhutan

The solution to this problem lies on producing durable wear-resistant coating on the turbine blades, so as to not only increase the original design life when used in OEM (original equipment manufacturing) but can also be coated several times during scheduled maintenance and repairs (M&R), thus extending the life of the turbine several times the original designed life. The published literatures on this particular problem indicate vast improvement in the life cycle of turbine runners with anti-wear overlays on the blades[5,6,10]. However, the nodal agencies like NHPC (National Hydro Power Corporation) in India and some of the overseas suppliers choose to conduct research in various alternatives. Based on their research findings, the field trials conducted so far have failed miserably. Hence, instead of developing antiwar coatings, the NHPC have signed an agreement recently (September, 2008) for *a new material of construction* with a national laboratory, to replace the currently used base material[11]. This being a long term project, and vested normally to a turbine manufacturer, hence the saga would likely to continue.

Renewable Energy (excluding big hydro)

The renewable energy, excepting big hydro plants, which have joined the mainstream production processes include, wind, biofuel, geothermal, solar PV and solar Thermal. The installed capacities in top six countries for five types of green resources in terms of gigawatts GW) are shown in table 12.4.[12,13] The total installed capacity in six top countries is around 80 GW, for the five green resources.

Table 12.4 Installed Generating Capacity of Renewable Energy; 2006 (Gigawatts)

Country	*Wind*	*Biomass*	*Geothermal*	*Solar PV*	*Solar Thermal*
Germany	22.5	2.0		3.0	
USA	11.0	7.0	3.5	0.5	0.5
Spain	12.0	0.5		0.1	
India	6.5	1.5			
Japan	1.5	3.5		0.5	1.2
China	2.7	1.9			

In 2006, overall renewable power capacity from all green resources, expanded from 160 GW in 2004, to 182 GW in 2006, and then to 240GW in 2007, excluding large hydropower. Global power capacity from new renewable energy sources (excluding large hydro) reached 280 GW in 2008 – a 16 percent rise from the 240 GW in 2007 and nearly three times the capacity of the United States nuclear sector[13].

The small scale power generation plants based on renewable resources face the difficulty of using transmission grids operated by big power generating plants based on conventional sources. Apart from competition high price of the power based on renewable resources is also a factor in this denial of access to grid. However renewable energy regulations introduced by various countries have resulted in granting access to transmission grids to power generated from renewable sources. For example, German law encourages investment by cross-subsiding renewable electricity fed into the grid[12]. America's renewable energy boom and the solution to grid sharing problem are due to state-laws requiring utilities to generate a certain share of power from renewable resources with generous tax incentive to abider[14].

Wind Power

In generating power from wind, the driving speed of airflow is used to run wind turbines. Modern wind turbines range from around 600 kW to up to 5 MW of rated power, although turbines, with rated output of 1.5–3 MW, have become the most common for commercial use. The power output of a turbine is a function of the cube of the wind speed. Therefore with the increase in wind speed, power output increases dramatically. Areas where winds are stronger and more constant, such as offshore and high altitude sites are preferred locations for wind farms. Wind power is based on renewable & sustainable resource and produces no greenhouse gases during operation.

Wind strengths near the Earth's surface vary and thus cannot guarantee continuous power and due to intermittent wind strengths, the capacity factor is 25%. It would mean that a typical 5 MW turbine in the EU would have an average output of 1.7 MW. It is best used in the context of a system that has significant reserve capacity such as hydro, or reserve load, such as a desalination plant, to mitigate the economic effects of resource variability.

Wind power is one of the fastest growing of the renewable energy technologies, though it currently provides less than 0.5% of global energy. Over the past decade, global installed maximum capacity increased from 11 GW in 2000 to 64.7 GW in 2006, grew by 28% to 95 GW by 2007, to 121 GW in 2008, another 28% growth—a trend that is projected to continue into the future (1,13).Windpower in USA grew 45% in 2007. Wind power accounts for around 1% electricity generated in USA.This figure is expected to rise to 15% by 2020 (14). It currently produces less than 1% of world-wide electricity use, but accounts for approximately 20% of electricity use in Denmark, 9% in Spain, and 7% in Germany.

Solar Energy

The total resources of all *fossil fuels* amount to about 0.4 YJ total, while the availability of *solar energy* is 3.8 YJ per year[1]. Solar constant is the average amount of incoming solar radiation per unit area on the outer surface of Earth's atmosphere is approximately 1,367 watts per square meter. The average incoming solar radiation or the solar irradiance, taking into account the half of the planet not receiving any solar radiation at all, is one fourth the solar constant or 342 W/m^2 (ref, Chapter3). Of all the total incoming sunlight that passes through the atmosphere annually, *51% is available at the Earth's surface*. If the 51% of solar energy amounting to 174.4 W per sq. meter absorbed by the Earth's surface can be utilized for power generation instead of currently used energy from burning fossil fuels, there would not be any global warming due to anthropogenic emissions.

Solar heating capacity has increased by 15 percent to 145 gigawatts-thermal (GWth), while biodiesel and ethanol production showed an increase of 34 percent[13].

The energy collected from solar beam, called solar energy, can be used in many ways for direct heating of materials.A 20 MW per sq.in solar flux needs a fraction of seconds to melt materials like silicon carbide with a melting temperature of 3630°C. The development of solar furnaces has made it possible to utilize direct sunlight as concentrated high energy beams for heat treatment and fusion processes of materials[6].The process is yet to be commercialised. The current practice is to utilise electricity or fossil fuels for the heat treatment and fusion of materials.

Two most widely used commercial processes to generate electricity are as follows:

i. *Solar Thermal or Concentrating Solar Power (CSP):* In this process the solar beam is concentrated by mirrors of different designs, and the concentrated beam is allowed to focus onto boiler tubes containing water. The concentrated heat source makes superheated steam in the boiler to drive a turbine generating electricity. Solar radiation used for heating covers almost entire spectrum range, including near ultraviolet (305nm) through visible (700nm) to near infra red (2500nm).Many materials absorbs visible radiation better than infra-red. Solar radiation can be concentrated to produce solar flux at peak energy of 16MW to 100MW per square inch by using different types and numbers of concentrators. Solar thermal concentrators are of different designs[13], including, (i) long curved mirrors, called parabolic troughs, to focus light on a tube of fluid running just above them. This is the one used in "Nevada Solar One' generating up to 64MW capacity; (ii) a large number of smaller mirrors to focus light on a tower in their midst.This system is used in a 11 MW Spanish plant; (iii) others using long flat mirrors and devices resembling satellite dishes.

The three major problems in solar thermal power generation, include the need for steam generating water in the favored water scarcity area like deserts, intermittent power generation depending on day light, and the non availability of power transmission grids set up by big power plants based mostly on fossil fuels[15,16]. The water scarcity in areas with plenty of sunshine like desert is a problem in CSP plants. A project to build a plant in Nevada needs 20% of the water available in the area. Another project in the Californian desert faces difficulty in appropriate water rights, since California water law prohibits use of potable water for cooling. Newer designs require less water by using air-cooling to convert the steam back into water. The water is then returned to the boiler in a closed process which is environmentally-friendly. Compared to conventional wet-cooling, this results in a 90 percent reduction in water usage. The proposed Ivanpah Solar Power Facility in south-eastern California will use this new design[17].

In order to make the electricity generation process continuous, it is essential to generate and store sufficient amount of heat during bright sunny days. Heat is transferred to a thermal storage medium in an insulated reservoir during the day, and withdrawn for power generation at night. Thermal storage media include pressurized steam, concrete, a variety of phase change materials, and molten salts such as sodium and potassium nitrate. Heat storage allows a solar thermal plant to produce electricity at night and on overcast days. Additionally, the utilization of the generator is higher which reduces cost. This allows the use of solar power for base load generation as well as peak power generation, with the potential of displacing both coal and natural gas fired power plants.While only 600 megawatts of solar thermal power is up and running in 2009, a 400 megawatts is under construction, and another 14,000 megawatts CST projects being developed.[18]

ii. *Solar Photovoltaic*[17]*:* Photovoltaic systems use conventional solar panels, where directly convert sunlight into energy using the principles of the photovoltaic effect. The photovoltaic effect takes advantage of the properties of semiconductor materials, with silicon being the primary material used in photovoltaic solar cells. When photons strike the solar cell, electrons in the semiconductor material are shaken loose, allowing them to flow as electricity. This electricity is direct current (DC), and can be directly used to charge batteries, or can be connected to an inverter to power alternating current (AC) components, or to be connected to the local electrical grid.

Traditional photovoltaic systems are based on silicon. Silicon ingots are sliced into wafers that are fabricated into cells. Cells are combined into modules, which are packaged into end-user systems. Silicon-based solar cells have efficiencies of approximately 14-19%. However, newer systems that use gallium arsenide, another semiconductor material, can be made into thinner and more flexible modules. These "thin film" modules can presently produce efficiencies up to 30%, but currently cost more to fabricate than traditional silicon-based modules.[17].

Photovoltaic production has been doubling every 2 years, increasing by an average of 48 percent each year since 2002, making it the world's fastest-growing energy technology. At the end of 2008, the cumulative global PV installations reached 15,200 MW[1,13]. The world's largest solar-photovoltaic plant with 46 MW capacities is in Portugal, and another 40MW in Germany, both built in 2008. The proposed larger plants have capacities ranging from 100 to 600 MW[16].

Biofuel[19]

Advanced technologies to convert biomass in the electricity, gaseous and liquid biofuels, and even hydrogen are in the process of development[19]. Some of the advanced technologies and the available commercial processes for conversion of biomass into biofuel are discussed in this section

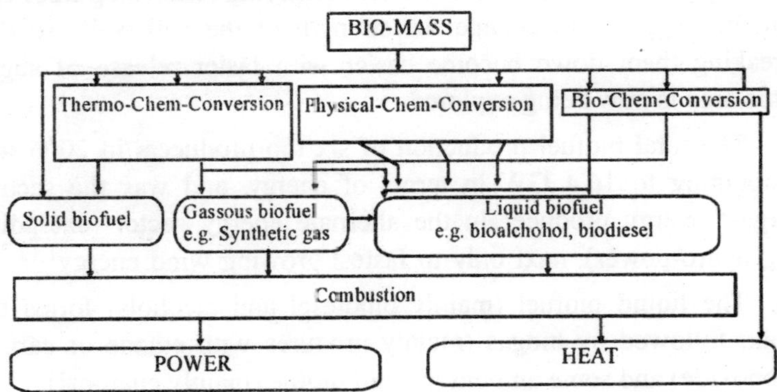

Fig.: Biomass Conversion Processes & Products

Plants use photosynthesis to grow and produce biomass or biomatter. Biomass, like other organic matters, such as fossil fuels is basically comprised of organic compounds of carbon and hydrogen. Therefore they can be used directly as fuel or can be converted to liquid or gas fuel. Biofuels are based on biomass, which is renewable and thus sustainable source of energy. The plants specifically grown for use as renewable and sustainable source for biofuels include corn & soybeans (in USA), flux seed and rapeseed (in Europe) sugarcane (Brazil), palm oil (Indonesia), and jatropha (India). Other sources of biofuel include biodegradable wastes, sewage, manure etc, which are converted to biogas by anaerobic digestion.

First generation biofuels, mainly bioethanol, biodiesel, and biogas which are produced from *food crops* (sugar or oil crops) and other food based feed stocks (*e.g.* food waste) are now commercially available, with almost 50 billion litres (39.5 million tons) of bioethanol and 5.4 billion litres (5.4 million tons) of biodiesel produced worldwide in 2006[19].

Second generation biofuels make use of a wider range of feedstocks, mainly *non-food crops*. For example, the whole plant biomass can be used or waste streams that are rich in lignin and cellulose, such as, wheat straws, grass, or wood.

Third generation biofuels make use of *biotechnological processes* to engineer the properties of plants so as to derive maximum amount of intermediate and /or final product with easier processing techniques. For example, plant scientists trying to develop trees that can be triggered to change the strength of the cell walls so that breaking them down become easier with faster release of sugar. This area is now being explored.

The total biofuel production of six top producers in 2006 was amounting to 16.4 GW in terms of energy, and was the second largest energy producer in the alternate energy sector (excluding big hydro-power), next only to fastest growing wind energy[1,12].

The liquid biofuel (mainly biodiesel and alcohols) forms the bulk, followed by biogas (mainly mixtures with ethane or carbon monoxide) and some amount in solid biofuel(mainly charcoal).

Biomass is converted to biofuels by thermo-chemical, physical-chemical and bio-chemical processes (tab.12.5)[19]. Thermo-chemical processes include, carbonization, gassification and pyrolysis. Pysical-chemical processes consist of pressing/extraction, and transesterification. Bio-chemical processes include alcoholic fermentation, anaerobic fermentation and composting.

Biofuels, such as, biodiesel and ethanol can be burned in internal combustion engine to generate motive power (thus to run car) or in boilers to generate electricity. Liquid biofuels are mostly used in the transport industry because of their lower carbon emissions than that of petrol and diesel

Liquid biofuel

Liquid biofuel constitutes the bulk of the biofuel produced by all the three process routes. All the products of physical-chemical processes, alcoholic fermentation products in bio-chemical processes, and part of gasification conversion products in thermo-chemical belong to liquid biofuel[19]. Liquid biofuel is usually either a *bioalcohol* such as ethanol or a *bio-oil* such as biodiesel and straight vegetable oil. Biodiesel can be used in modern diesel vehicles with little or no modification to the engine and can be made from waste, virgin vegetable, animal oil and fats (lipids). Virgin vegetable oils can be used in modified diesel engines. The use of biodiesel reduces emission of carbon dioxide and other hydrocarbons by 20 to 40%.

In some areas corn, cornstalks, sugar beets, sugar cane, and switchgrasses are grown specifically to produce ethanol (also known as grain alcohol) a liquid which can be used in internal combustion engines. Ethanol is being phased into the current energy infrastructure. E85 is a fuel composed of 85% ethanol and 15% gasoline that is sold to consumers. Biobutanol is being developed as an alternative to bioethanol.

Synthetic Petrol

Syngas can be directly used in internal combustion engines, or to produce methanol and hydrogen, or converted via the Fischer-Tropsch (FT) process into synthetic fuel. The principal purpose of this FT process is to produce a synthetic petroleum oil or synthetic petroleum substitute, such as synthetic lubrication oil for running trucks, cars, and some aircraft engines. *The combination of biomass gasification (BG) and FT synthesis is considered by some as very promising route to produce renewable transportation fuels.* Using natural gas as a feedstock, the ultra-clean, low sulfur fuel has been tested extensively by the US Department of Energy, the Department of Transportation, and most recently, with US *Air Force to develop a synthetic jet fuel blend*[19].

Pitfall in biofuels making

The conversion of biomass into biofuel in the form of liquid or gas involves consumption of energy. In some cases the energy consumed in the conversion process can be more than the saving in energy by substituting fossil fuels. In the case of corn-ethanol fuel making, if all the corn produced in USA were converted to ethanol this would save only 12% of gasoline demand. However, one gallon of corn ethanol requires four-fifth of a gallon of fossil fuels and 1,700 gallons of water to produce. The net advantage tends to be negative[20].

Use of liquid biofuel

Brazil has one of the largest renewable energy programs in the world, involving production of ethanol fuel from sugar cane, and ethanol now provides 18 percent of the country's automotive fuel. As a result of this, together with the exploitation of domestic deep water oil sources, Brazil, has recently reached complete self-sufficiency in oil. Most cars on the road today in the U S, run on blends of up to 10% ethanol, and some up to 85% ethanol (E85).

Solid biomass

Solid biomass is mostly used directly as a combustible fuel, producing 10-20 MJ/kg of heat. The sources include *wood fuel*, the biogenic portion of municipal solid waste, or the unused portion of field crops. Two billion people currently cook every day, and heat their homes in the winter by burning biomass, which is a major contributor to man-made climate change. The black soot that is being carried from Asia to polar ice caps is causing them to melt faster in the summer. Wood and its byproducts can now be converted into biofuels such as *woodgas*, methanol or ethanol fuel. Further development may be required to make these methods affordable and practical.

White Coal from Biomass in India

The white coal is produced from agricultural waste, like mustard sticks and sesame seed oilcake. After mixing and grinding, the powder is shaped as briquettes, and sold as white coal, for use in boilers and brick kilns. The product has become popular in Rajasthan in India where more then 50 units are producing the biofuel with low level of carbon emission. During last three to four years, an investment of around 200 million rupees (approximately over 4 million USD) has been made in the white coal production from the bio waste[21].

Biogas[19]

Biogas can easily be produced from current waste streams, such as, paper production, sugar production, sewage & animal waste.The various waste streams have to be slurried together and allowed to naturally ferment, producing methane gas. This can be done by converting current sewage plants into biogas plants. When a biogas plant has extracted all the methane, the remains work better as fertilizer than the original biomass. Alternatively biogas can be produced via advanced waste processing systems such as mechanical biological treatment. These systems recover the recyclable elements of household waste and process the biodegradable fraction in anaerobic digesters. Renewable natural gas is a biogas which has been upgraded to a quality similar to natural gas. By upgrading the quality to that of natural gas, it becomes possible to distribute the gas to the mass market via gas grid.

Advanced Gasification

Gasification process is used to convert any type of organic material, such as coal, petroleum, or more recently biomass or organic waste into carbon monoxide and hydrogen by controlled combustion with oxygen.The gas mixture produced is called *synthetic gas or syngas* and is itself a fuel. Gasification is a very efficient method for extracting energy from many different types of organic materials, and also has applications as a clean waste disposal technique. The gasification may be an important technology for renewable. In particular biomass gasification can be carbon neutral. Gasification relies on chemical processes at elevated temperatures >700°C, which distinguishes it from biological processes such *as anaerobic digestion* that produces *biogas*. Liquids are more portable because they have high energy density, and they can be pumped, which makes handling easier. This is why most transportation fuels are liquids.

Biofuels from Seaweeds by Aquaculture[22,23]

Seaweeds have the potential of not only countering global warming but also providing bio-fuels to cater the growing energy needs. Large-scale cultivation of biofuels on land has serious environmental costs, including deforestation, water use and generation of greenhouse gases. In recent years, the use of agriculture land & resources has led to growing increase in food scarcity and corresponding price increase. These could be avoided by seaweed cultivation.

Technology for aqua- or sea- or mari-culture[22,23]

Seaweeds are *marine plants* and *protists* (a diverse group of eukaryotes that cannot be classified as animals, plants, or fungi) belonging to the category of *benthic algae* and often found in the seashore *biome.*The main components of seaweed are fucoidan and alginic acid. While an enzyme for breaking down fucoidan to sugar has already been discovered, the scientists are looking for an enzyme that breaks down alginic acid. They are also looking at the possibility of using genetic modification technology. The sequences in making ethanol from seaweeds are as follows:

Seaweed (fucoidan +alginic acid) → Fucoidan + Enzyme → Sugar → Ethanol

Proposed Giant Plant in Japan

Tokyo University, along with Mitsubishi Heavy Industries and several other private-sector firms envision a 10,000 square kilometer (3,860 square mile) seaweed farm at Yamatotai, a shallow fishing area in the middle of the Sea of Japan, to cultivate seaweed, which grows rapidly, a floating reactor for converting seaweed to ethanol at sea, and to transport by tankers.They claim a farm of this scale could produce about 20 million kiloliters (5.3 billion gallons) of bioethanol per year, which is equivalent to one-third the 60 million kiloliters (16 billion gallons) of gasoline that Japan consumes each year.

The researchers claim that in addition to serving as a source of fuel, the seaweed would help clean up the Sea of Japan, by removing some of the excess nutrient salts that flow into the sea from the surrounding land masses. Along with Japan, seaweed farming has been established in Costa Rica, with R&D inputs from World Bank funded Sea Gardens Project at the University of Costa Rica.

Less than three per cent of world's oceans or about 20% of agricultural land used in agriculture would be needed to fully substitute for fossil fuel. The former is a much more superior alternative than the later. One concern of harvesting naturally occurring seafood could have comparable GHG emission effects due to habitat loss or fragmentation.The effect may be as good as that of large scale deforestation.

Seaweeds from Waste Water

Growing large seaweed fields for energy by using wastewater nutrients could be economically sound, for millions of tones of untreated wastewater are dumped daily into seas and seaweeds help clean it up. This idea has been tested successfully using human waste water in experiments in USA.

Biorefinery[19]

A *biorefinery* is a facility that integrates biomass conversion processes and equipment to produce fuels, power, and value-added chemicals from biomass. Similar to petroleum refinery, a biorefinery produces multiple products such as, chemicals, biodiesel or bioethanol, and simultaneously *generating electricity* and *process heat*, through *combined heat and power* (CHP) technology, for its own use and perhaps enough for sale of electricity to the local utility.Although some facilities exist that can be called bio-refineries, the bio-refinery has yet to be fully realized.

According to the International Energy Agency, amongst the new bioenergy (biofuel) technologies, most promising one is cellulosic ethanol biorefineries. Cellulosic ethanol can be made from inedible cellulose fibers that form the stems and branches of plants. Crop residues (such as corn stalks, wheat straw and rice straw), wood waste, and municipal solid waste are potential sources of cellulosic biomass. Dedicated energy crops, such as switchgrass, as sources for cellulose can be sustainably produced in many regions.

Royal Society on Biofuel[24]

Theoretically biofuel produces lower carbon dioxide, which would get absorbed by the growing plants and the recycling process continues. Farmers hope to make more money by growing crops for biofuels than other agricultural products.

The Royal Society, UK's national science academy, based on the analyses of a wide range of commercially available biofuels have published a report, in which it was concluded that some *biofuels may in effect cause more global warming than petrol due to emissions of greenhouse gases from fertilizers* and processing. Royal Society argues for a worldwide strict certification system, similar to that used for eco-friendly wood, to let consumers know just how green is the biofuel they would be using.

Development of such a system would require exhaustive analysis of every step of a firm's supply chain, a difficult, if not an impossibie task, in view of fiercely protected indigenous biofuel industries in USA and Brazil.

Geothermal Energy[25]

Geothermal energy is obtained by tapping the heat of the earth itself, usually from kilometers deep into the Earth's crust. It is expensive to build a power station but operating costs are low resulting in low energy costs for suitable sites. It should be stressed that the geothermal resource is not strictly renewable in the same sense as the hydro resource. However, the International Energy Agency considers geothermal power as renewable.It estimates that Iceland's geothermal energy could provide 1700 MW for over 100 years, compared to the current production of 140 MW. Geothermal is the first generation sustainable energy source.

Three types of power plants are used to generate power from geothermal energy: dry steam, flash, and binary. Dry steam plants take steam out of fractures in the ground and use it to directly drive a turbine that spins a generator. Flash plants take hot water, usually at temperatures over 200 °C, out of the ground, and allows it to boil as it rises to the surface then separates the steam phase in steam/water separators and then runs the steam through a turbine. In binary plants, the hot water flows through heat exchangers, boiling an organic fluid that spins the turbine. The condensed steam and remaining geothermal fluid from all three types of plants are injected back into the hot rock to pick up more heat.

Alcoa, US-based aluminum giant, is evaluating the feasibility of building the world's first big geothermal power plant in Iceland for running a proposed aluminum production plant.[25]

Nuclear Power[26,27]

Fission of uranium atom by accelerated neutron generates large quantity of energy. Thorium can also be used. Fission power's long-term sustainability depends on the amount of uranium and thorium that are available to be mined. It is said that nuclear has the potential to be sustainable. However, this is often qualified with the argument that there are serious challenges with respect to nuclear hazards that must be dealt with before it can drastically increase its role.Nuclear fission is generally not regarded as renewable, as indicated by the U.S. DOE on the website[1]. Both fission and yet to be developed commercially fusion process, create radioactive waste in the form of activated structural material, which is one of the sustainability issues.

Currently, there are 440 commercial power generating nuclear reactors operating in 30 different countries. The total installed capacity is 372,000 Mw, which supplies 16% of global electricity consumption[3]. Proponents, claim that nuclear power is at least as environmentally friendly as traditional sources of renewable energy, making it part of the solution to global warming and the world's growing need for energy.

Due to high energy cost and environmental problem, some of the major nuclear energy generating countries, like USA and UK have stalled the further expansion projects[26,27]. US have 104 nuclear power reactors at 65 power plants, and no newer one under construction[25]. After 1995, no new nuclear power plant has been commissioned in U.K. In 2020 nuclear energy would provide only 7% of the requirement, from existing 18%[27]. The nuclear establishments in UK are opting for improving lifetime of efficiency and updating nuclear waste management rather than going for new plants[26]. Large part of electricity supply in Germany is from nuclear power plant, but they are rapidly making stride to replace the need by green renewable resources[12]. Aim is to maintain existing plants and keep the show on road. However in Europe, a total of 197 European nuclear power plants are in operation and another 13 under construction in five countries in Europe.

The recent revival of interest in nuclear energy is more due to spiraling oil and gas price combined with increasing demand rather than environmental concern. Also the economics on nuclear power has changed due to high fossil fuel price, making the cost to consumer similar to conventional power. Thirdly new designs of future nuclear power plant are being explored to make it more techno-economically viable[26]. So far as fusion energy is concerned it is still in laboratory, and commercial production may take another 30 years or more.

Mini or Micro-generation Facilities

The 'climate change and sustainable energy act 2006' in UK aims to boost the number of electricity microgeneration installations in the country, so helping to cut carbon emissions and reduce fuel poverty. For the purposes of the Act, microgeneration technologies include, biomass, biofuels, fuel cell, photovoltaics, water (including wave and tidal), wind power, solar power,geothermal sources and combined heat and power systems.

Some Third-generation technologies

Hot-dry-rock power unit uses geothermal source, where the very high temperature of rocks just a few kilometers below ground, is used to produce steam and run a turbine to generate electricity.

Ocean energy[28] is another third-generation technology. Portugal has the world's first commercial ***wave farm***, the *Aguçadora Wave Park*, under construction in 2007, shall generate 2.25 MW intially and based on the success, there would be further expansion to a generation capacity of 525 MW. A wave farm in the world's first commercial ***tidal power*** station (2007) in the narrows of Strangford Lough in Ireland, shall generate 1.2 megawatt underwater tidal electricity.[27]

Solar power panels that use *nanotechnology*[29] can create circuits out of individual silicon molecules, may cost half as much as traditional photovoltaic cells. Nanosolar, USA has built a factory for nanotechnology thin-film solar panels and reported to make roll-print solar cells that require only 1/100th as thick an absorber as a silicon-wafer cell. Company's plant has a planned production capacity of 430 megawatts peak power of solar cells per year[28].

Renewable energy and ***energy efficiency*** are sometimes said to be the "twin pillars" of sustainable energy policy. Both resources must be developed in order to stabilize and reduce carbon dioxide emissions. Renewable energy (and energy efficiency) is no longer niche sectors that are promoted only by governments and environmentalists. The increased levels of investment and the fact that much of the capital is coming from more conventional financial sectors suggest that sustainable energy options are now becoming mainstream.

India is *rich in green energy resources* – clean, renewable and sustainable energy resources like, solar, wind, water and biofuels (jatropha, sugarcane) - which promise significant future potential.The efforts to develop energy from these resources shall pay rich dividend both financially and creating a truly green country.

Summary

i. The natural sustainable alternate sources of energy comprise of water, wind and sunbeams. These are zero-emission sources. Other sources include reduced carbon organic materials, such as, biomass.

ii. Till date the fossil fuels based power generating plants produce most of the electricity. Technological developments have made possible the development of high efficiency, high capacity super critical boilers & generation units, such as, Ultra Mega, of 4000 megawatts or above, with much lower carbon emissions than conventional processes.

iii. Amongst the alternative, the hydroelectric plants have been used as a major source of commercial production of electricity. Hydroelectric is the clean, green, and zero emission projects. Countries like Norway, Brazil and Venezuela have most of the power requirements produced by hydro electric plants

iv. Some countries have used nuclear power plants to cater for a large portion of the electrical power requirements. Quite a few major operators, like UK and USA, have not opted for further growth in this area.

v. Alternatives, such as wind power and solar energy with high growth rate have already joined the mainstream. Boifuel is already playing a major role with land based resources. Biofuel from seaweeds has great potential.

vi. Renewable clean energy and energy efficiency- are keys to combat global warming.

REFERENCES

1. U.S. Energy Information Administration, International Energy Outlook 2009, Report #:DOE/EIA-0484 (2009), May 27, 2009. http://www.eia.doe.gov/ oiaf/ieo /electricity.html.Table12.1 gives energy consumption in 2004

1a. International Energy Agency (IEA) report on CO2 emission, Paris, *Times of India*, 31st May,2011

2. Tata Ultra Mega; IFC website: ifc.org/ifcext/spiwebsite1.nsf/

3. Carbon Monitoring for Action (CARMA) of Center for Global Development, Washington, USA, article on 'China Passes U.S., Leads World in Power Sector Carbon Emissions 'August 27, 2008, http://www.cgdev.org/content/article/detail/16578/

4. R.Chattopadhyay: Advanced Thermally Assisted Surface Engineering Processes, Kluwer Academic Publishers, MA, USA, 2004. (Now Springer, NY)

5. R. Chattopadhyay: Surface Wear – Analysis, Treatment, and Prevention, ASM International, Materials Park, OH, USA, 2001

6. R. Chattopadhyay, invited speaker, Reconditioning of Thermal & Hydro Power Plants-Indian Experience: Indo-French Seminar on Efficient Management of Power Plants,16-17feb,1995, Taj Hotel,Bombay

7. Improving the Efficiency of Coal-fired Power Plants Reduces CO_2 Emissions, World Coal Institute (http://www.worldcoal.org/coal-the-environment/coal-use-the-environment/improving-efficiencies)

7a Supercritical pressure coal fired thermal power plants, http://www.hitachi.com/environment/showcase/solution/energy/thermal_power.html

8. Zhen Fan, Steve Goidich, Archie Robertson and Robert Roche, *Performance and economics of ultra supercritical pressure CFB boiler power Tplant*, Foster Wheeler North America Corp., Livingston, NJ 07039,http: //www.fwc.com/publications/tech_papers/files/TP_CFB_07_04.pdf

9. International Energy Agency, Key World Statistics, 2007, Producers of Hydro Electricity; http://www.iea.org/textbase/nppdf/free /2007/key_stats_2007.pdf

10. R. Chattopadhyay; High Silt Wear of Hydro-Turbine Runners: Paper presented at 9th. Int. Conf.on 'Wear of Materials'; San Francisco, USA, 13 – 16 April, 1993, published in the proceedings by Elsevier Publication

11. National perspective plan on R&D in the power sector, NHPC; http://www .nhpcindia. com/ English/Scripts/RD_Thrustareas.aspx

12. Renewable energy policy network, REN21, http://www.ren21.net globalstatusreport/ g2009.asp

13. German lessons, The Economist, 5th April, 2008
14. Solar Energy, The power of Concentration, *The Economist*, February 23rd, 2006, p71.
15. Todd Weedy, *Alternate Energy Projects Stumble on a Need for Water*, The New York Times, September 29, 2009.
16. P. Manning, With green power comes great responsibility. Sydney Morning Herald., 10 October 2009,http://www.smh.com.au/business/with-green-power-comes-great-responsibility-20091009-gqvt.html.
17. Generating Power from Solar Energy - Using Photovoltaic Process Mar 31,2008,http://engineering.suite101.com/article.cfm/generating_power_ from _ solar_energy#ixzz0aJIIWbaS
18. *The Environment*, p. 104, The World in 2009 issue, The Economist.
19. Background Paper on Biofuel Production Technologies, Working document, November, 2007, prepared by Sergey Zinoviev, Sivasamy Arumugam, and Stanislav Miertus. International Centre for Science and High Technology,UNIDO,http://www.biofuels-dubrovnik.org/downloads Background% 20Paper% 20biofuels% 20CEE% 20UNIDO.pdf
20. Micheal Shank, letters to editor, *The Economist*, 2nd February, 2008
21. White coal from biomass Jan 13, 2010 Yahoo. News. (http://in.news.yahoo.com/139/20100113/824/tnl-white-coal-becomes-a-popular-fuel-so.html)
22. Scidev.Net, *Hindustan Times*, Mumbai, India, and June 10, 2008
23 Ricardo Radulovich, University of Costa Rica:Seaweeds, the coming revolution http://www.maricultura.net/wp-content/uploads/2008/06/seaweeds-the-coming-revolution.pdf
24. Current status of biofuel in UK, p. 54-5, *The Economist*, Jan 19th, 2008.
25. Lisa A. Swenarski de Herrera, *Span magazine*, March-April, 2008,p14-17 geothermal from bauxite.
26. Margaret W Hunt, Energy Independence, editorial, Advanced Material & Processes, ASM-International, July, 2008.
27.. *Current status of nuclear power in UK*, Materials World, UK Jan, 2009
28. International Energy Agency, Implementing Agreement on Ocean Energy Systems (IEA-OES), Annual Report 2007
29. Nanosolar, CA, USA homepage, http://www.nanosolar.com

14

GHG Management in Manufacturing Industries

The list of energy intensive, large primary manufacturing industries include chemical, iron and steel, aluminum, cement, paper and pulp, petrochemicals and other minerals and metals. The less energy intensive sectors or light industries include the rest, such as, manufacturing of food, beverages, tobacco; textiles; wood and wood products; printing and publishing; production of fine chemicals; and the metal processing industry (including automobiles, appliances, and electronics).The emission management in five large energy intensive industries (chemical, iron & steel, aluminum, cement, and paper & pulp) is to be discussed in this chapter.

Non-carbon dioxide GHGs emitted from the manufacturing sector include nitrous oxide (N_2O), hydrofluorocarbons (HFCs), perfluorocarbons (PFCs) and sulphur hexafluoride (SF_6). The use of substitutes for ozone depleting gases has been found to be an effective step to reduce fluorocarcarbon and sulphur hexafluoride.GHG management in manufacturing industries is thus mainly restricted to minimize carbon dioxide emission and in some industries nitrous oxide.

Emissions of carbon dioxide are still the most dominant contribution of manufacturing industry to total greenhouse gas emission. In emission data in manufacturing industries include all direct emissions plus indirect equivalent emissions due to electricity consumption. Nearly a third of the world's energy consumption and 36% of carbon dioxide (CO_2) emissions are attributable to manufacturing industries. The carbon dioxide emission (36%) figure in manufacturing industries follows the energy consumption pattern (33.3%)[ref.1]. The large primary materials industries, *i.e.*, *chemical, petrochemicals, iron and steel, cement, paper and pulp, and other minerals and metals*, account for more than two-thirds of these figures, *i.e.* around a quarter (24%) of total global energy consumption and a quarter of total global carbon dioxide emission. Of metals, iron and steel and aluminum production are very high energy consuming processes and also big emitters.

Overall, industrial energy use has been growing strongly in recent decades. The rate of growth varies significantly between sub-sectors. For example, chemicals and petrochemicals, which are the heaviest industrial energy users, doubled their energy and feedstock demand between 1971 and 2004, whereas energy consumption for iron and steel has been relatively stable.Aluminium industry has been able to reduce emissions considerably due mainly to their R&D efforts,

In recent times, major part of the industrial growth has been in emerging economies. Today, China is the world's largest producer of iron and steel, ammonia and cement and the worst emitters, but not a signatory to Kyoto Protocol.

Need to Conserve Non-renewable Resources to Reduce Emission

To cater for the needs of ever growing population, economic growth at a reasonable rate is essential. Economic growth is expressed in terms of percentage increase in annual GDP (gross domestic products).To produce the increased quantities of the products it is necessary to consume more natural resources, as materials and energy inputs.

Steady growth is, by its nature, an exponential function. A quantity that grows according to an exponential function exhibits a doubling in size at a regular time interval (called the *doubling time*). With a steady growth rate of 5% per year, the consumption of a *non-renewable resource* (like coal, petrol, minerals, metals) would be double in approximately 14 years. After another 14 years the rate will have quadrupled. After a century of 5% annual growth, the resource will be consumed at a rate 130 times the original rate. The rapid depletion of non-renewable resources would make the economic growth unsustainable. The manufacturing industries need to reduce consumption of non-renewable resources for both material and energy needs to make the industrial growth at a sustainable rate.An effective step in this direction would cause substantial reduction in emissions.

Emission Management by Industries

In order to limit emissions, industries have adopted appropriate measures to suit their specific requirements. Some of these measures include followings:

i. Developing energy efficient processing systems. The innovative changes in the processing system can lead to higher productivity per unit of power consumption. NEDO (Japan) has published a comprehensive report on global warming counter measures through Japanese technologies for energy savings/GHG emission reductions[2]. The report covers technologies developed by NEDO for almost all the industries for saving basically energy. Report also contains the reduction in GHG emissions as achieved by industries. Emission efficiency is indicated by the volume of product shipped per unit of equivalent carbon emission. In Japan, electronic industries use JEITA (Japan Electronic and Information Technology Industries) common index, which is the value of CO_2 emission against net production (monetary value), adjusted using the corporate goods price index announced by the Bank of Japan. A fall in this figure indicates that a given product quantity (monetary value) is produced using less energy (see Chapter 19 on CSR, p.226). Industries have, in general, improved energy and emission control efficiencies by producing more with less energy consumption.

ii. Reduction in direct emission by increasing the proportion of green inputs, carbon capturing and altering the processing techniques.

iii. Improving life cycles of processing equipments shall lead to increase productivity, less downtime, lower carbon footprints of the equipments, savings in materials and energy due to less requirement of new components etc[3,4].

Sectorial Approach

All major sectors of industries have formed common platform of the companies belonging to the sector in their efforts to control emissions. Quite a few of these platforms are global. Sectoral policy ideas such as this could apply to companies not just in the steel sector but in other industries that produce a lot of carbon dioxide – including electricity generation, cement, chemicals and aluminum. One idea is that businesses active in such areas could, in a post-2012 governmental regime (post Kyoto), be awarded carbon dioxide "permits" by governments on the basis of how proficient they are at cutting greenhouse emissions per unit of economic output.

a. Electricity Generation

Power plants account for 40 percent of U.S. greenhouse gas emissions and 25 percent of the world. China, South Africa and India host the world's five dirtiest utility companies in terms of global warming pollution, while a single Southern Co plant in Juliette, Ga., USA emits more annually than Brazil's entire power sector. Emission management in power plants has been extensively covered in chapter 12.

b. Chemical industry[1]

Two major emitters of *greenhouse gases in chemical manufacturing industries include ammonia and nitric acid.* While ammonia production process leads to generation of carbon dioxide, the production of nitric acid can cause nitrous oxide emission. A number of technologies for nitrous oxide mitigation during nitric acid production are available. China is one of the leading producers of ammonia.

Ammonia (NH3) is a major industrial chemical and the most important nitrogenous material produced. Ammonia gas is used in producing fertilizer; heat treating and paper pulping; manufacturing of nitric acid, nitrates, nitric acid ester, nitro compound, explosives of various types, and as a refrigerant. Amines, amides, and miscellaneous other organic compounds, such as urea, are made from ammonia.The production of ammonia represents a significant non-energy industrial source of CO_2 emissions. The primary release of CO_2 at plants using the natural gas catalytic steam reforming process occurs during regeneration of the CO_2 scrubbing solution with lesser emissions resulting from condensate stripping.

During the production of nitric acid (HNO_3), nitrous oxide (N_2O) is generated as an unintended by-product of the high temperature catalytic oxidation of ammonia (NH_3) and is a significant source of atmospheric N_2O. If not abated, this would be the major source of N_2O emissions in the chemical industry. A number of technologies for N_2O mitigation during nitric acid manufacture have been developed in recent years. Examples include, option involving direct catalytic decomposition right after the platinum gauzes, and a full-scale catalyst decomposition option. Nylon production is also responsible for nitrous oxide emission through its requirement for adipic acid production. However, improvements in the removal of nitrous oxide before emissions to the atmosphere, such as by thermal decomposition, can prevent potential emissions form this source by over 90 percent.[5]

According to ACC (American Chemistry Council (ACC),

- EPA's (US Environmental Protection Agency) own measures, the chemical industry has maintained N_20 emissions from 18.9 Tg CO_2 eqiuvalent (Tetra gram of CO_2 equivalent) to 19Tg CO_2 equivalent in 2008, even as the chemical production has increased[5]
- Emissions of six criteria air pollutants are down over 43% nationwide even as the economy, GDP and population have expanded.

- Industrial discharges to the nation's waterways are but a small fraction of the remaining problems that are now driven by urban and agricultural runoff.

These advances have been achieved, in part, as a result of the development of new technologies and continued investment in new equipment, processes and procedure CIA (Chemical Industries Association, UK) reports that the chemical sector in UK has exceeded its own 2004 climate change agreement target. The sector has improved its energy efficiency by 19.5 per cent since 1998, equivalent to an annual saving of around 3.5 million tonnes of carbon dioxide emissions.

c. Paper and Pulp Industry[1,6,7]

Paper industry accounts for ~70% of industrial (nonutility) carbon dioxide emissions[1]. The figure roughly amounts to 7 million tons of the 10 million tons of industry-wide carbon dioxide emissions. Carbon dioxide reduction potential from industry sector largely tied to paper industry

Emissions Characterization[6]

On the average during 2001-2005 period, (i) 31% of paper industry CO_2 emissions correspond to carbon neutral, (ii) 55% of paper industry CO_2 emissions result from coal combustion, and (iii) 14% of paper industry CO_2 emissions result from gas, oil combustion. Thus the highest emission (55%) results from coal combustion, hence greatest potential for reduction is to substitute coal by low emission fuels.

Technologies to improve energy efficiency and reduce CO_2 emissions include

i. *Switching over from coal to natural gas:* The CO_2 emission from natural gas is around 56% of coal emissions (adjusted for efficiency). However natural gas is ~3.7 times more expensive than coal and thus not a cost effective solution to paper industry.

ii. *Improvement potential in chemical pulping: Black liquid gasification:* The paper and pulp industry is to play a leading role in the development of second-generation biofuels, such as, gasifying & refining so-called "black liquor" – the oily liquid residue produced in pulping wood to produce paper – to produce both bio-synthesis gas and liquid fuel. The average U.S. integrated pulp and paper mill has a thermal demand of approximately 40% fossil fuel and ~60% biomass, which is largely met from combustion of black liquor. With a biorefinery, there is no longer an input for fossil fuel-based energy since the pulp and paper facilities run on recovered heat.The outputs include pulp and paper, plus one or more 'green' fuels or chemicals. Power input will be an option determined by its cost versus the value of other output streams.

iii. *Development in mechanical pulping:* The efforts to develop mechanical pulping as a part of advanced paper making technologies have high potential to reduce CO_2 emissions

iv. *Energy conversion: Increasing the use of CHP systems:* Recent studies have identified combined production of heat and power (CHP) as one of the most important technologies for improving energy efficiency and reducing carbon emissions in the US. CHP is especially attractive in industries with constant steam loads and those that generate byproduct fuels. Chemicals and pulp/paper industries are the two largest industries dominating the CHP market. Combined CHP capacity in these two industries in 1994 was 24.2 GW — 55% of the total industrial CHP capacity[4]. Currently, CHP capacity in both industries has been realized mostly at the sites with high steam loads. However, significant potential still exists at the remaining sites. The 55% of CHP generation in these two industries would reduce carbon emissions equivalent to 44% of the present carbon emissions in these industries[4]. Also most of the carbon emission reductions can be achieved at negative costs. A barrier for CHP in India is the relatively low plant load factor of 0.534, indicating fluctuations in heat or electricity demand necessitating special care to be taken in plant design and demand side energy management[7].

d. Metal Production

i. Iron & Steel[8,9]

The integrated iron and steel plant is one of the world's highest emitters of greenhouse gases, generating 4 per cent of global carbon dioxide emissions yearly. Iron and steel industries make use of a large quantity of coal, good quantities of gas or oil, and some quantities of limestone - all of them generate carbon dioxide. Also the industry needs captive power plant based on fossil fuels.

The production of steel at an integrated iron and steel plant is accomplished using several interrelated processes. The major processes are (1) coke production, (2) sinter production, (3) iron production, (4) raw steel production, (5) ladle metallurgy, (6) continuous casting, (7) hot and cold rolling, and (8) finished product preparation. The operations for secondary steelmaking, where ferrous scrap is recycled by melting and refining in electric arc furnaces (EAFs) include (4) through (8) above.[8]

The GHG emissions are generated as (1) process emissions, in which raw materials and combustion both may contribute to CO_2 emissions; (2) emissions from combustion sources alone; and (3) indirect emissions from consumption of electricity (primarily in EAFs and in finishing operations such as rolling mills at both integrated and EAF plants). The major processes for direct GHG sources[1] include the sinter plant; non-recovery coke oven battery combustion stack;coke pushing; basic oxygen furnace (BOF) exhaust; and EAF exhaust.The primary combustion sources[2] of GHGs include by-product recovery coke oven battery combustion stack; blast furnace stove; boiler; process heater; reheat furnace; flame-suppression system; annealing furnace;flare; ladle reheater; and other miscellaneous combustion sources.

The primary sources of GHG emissions are blast furnace stoves (43 percent), miscellaneous combustion sources burning natural gas and process gases (30 percent), other process units (15 percent), and indirect emissions from electricity usage (12 percent). For EAF steelmaking, the primary sources of GHG emissions include indirect emissions from electricity usage (50 percent), combustion of natural gas in miscellaneous combustion units (40 percent) and steel production in the EAF (10 percent)[8]. Some basic facts on emission & its control in iron & steel industry[9]:

- The iron and steel industry accounts for about 19% of energy use and about a quarter of direct CO_2 emissions from the industry sector. The CO_2 relevance is not linear to power consumption but quite high due to a large share of coal for coke production, coke for producing molten iron and the coal for captive power plant.
- The iron and steel industry has achieved significant efficiency improvements in the past twenty-five years. Increased recycling and higher efficiency of energy and materials use have played an important role in this positive development.
- Iron and steel has a complex industrial structure, but only a limited number of processes are applied worldwide. A large share of the differences in energy intensities and CO_2 emissions on a plant and country level can be explained by variations in the quality of the resources, the processes used and the overall efficiency.

Plant Efficiency in Iron and Steel Industry[(9)]*:* The efficiency is closely linked to several elements including *technology, plant size and quality of raw materials.* This partly explains why the average efficiency of the iron and steel industries in China, India, Ukraine and the Russian Federation are lower than those in OECD countries. *These four countries account for nearly half of global iron production and more than half of global CO_2 emissions from iron and steel production.* Outdated technologies, *e.g.*, open hearth furnaces are still in use in Ukraine & Russia. In India, new, but energy inefficient, technologies such as coal-based direct reduced iron production play an important role. These technologies can take advantage of the local low-quality resources and can be developed on a small scale, but they carry a heavy environmental burden. Raw material constraints, such as poor quality of coking coal also lead to energy inefficiency. The slow updating of technology is another constraint. In China, low energy efficiency is mainly due to a high share of small-scale blast furnaces, limited or inefficient use of residual gases and low quality ore.

Waste energy recovery in the iron and steel industry: This tends to be more prevalent in countries with high energy prices, where the waste heat is used for power generation. This includes technology options such as coke dry quenching (CDQ) and top-pressure turbines. CDQ also improves the coke quality, compared to conventional wet quenching technology.

Primary energy savings potential[1]*:* The identified primary energy savings potential is about 2.3 to 2.9 EJ per year through energy efficiency improvements, *e.g.* in blast furnace systems and use of best available technology. Other options, for which only qualitative data are available, and the complete recovery of used steel can raise the potential to about 5 EJ per year. The full range of CO_2 emissions reductions is estimated to be 220 to 360 Mt CO_2 per year. [* One exajoule (EJ) equals 1018 joules or 23.9 Mt CO_2.]

Global Sectorial Approach in Steel Industry: The steel industry has endorsed a global approach as the best way for steel to help address climate change. The International Iron and Steel Institute (IISI) approved the next stage in the establishment of a Global Sectorial Approach for steel. The organization will collect and report the carbon dioxide emissions data of steel plants in all the major steel producing countries. Establishment of the data on a common and consistent basis is the starting point for the setting of commitments post-2012 on a national or regional basis.

The steel industry in North America, Western Europe and Japan has reduced energy consumption per unit of production by 49 percent in the last 25 years, according to an IISI statement. The steel industry now accounts for three percent to four percent of global man-made greenhouse gas emissions. Over 90 percent of steel industry emissions come from iron production in nine countries or regions: Brazil, China, EU-27, India, Japan, Korea, Russia, Ukraine and the USA. In a study of global sectoral approaches (CCAP,

2010), CCAP investigated a transnational approach in which all countries face similar benchmarks, a sectoral Clean Development Mechanism (CDM) approach emphasizing carbon credits, and a bottom-up approach envisaging 8 of 69 financial and technology assistance from advanced economies to support ambitious no-lose crediting baselines in developing countries. This study was supported by the Competitiveness and Innovation Framework Programme of the European Commission, and the study's objective was to help move beyond voluntary actions and facilitate participation by developing countries in international climate change actions.[8]

Cap and trade regional policies such as those currently used in the EU are not effective in reducing carbon dioxide emissions, said Philippe Varin, IISI Executive Committee Member and CEO, Corus. "Constraining production from the best emission performing plants is not the solution for a globally competitive industry such as steel. An effective approach for the steel industry requires the participation of all major steel producing countries and a focus on improving emissions per unit of production."[1]

Energy Efficiency Improvements (8): The iron and steel industry is energy intensive; consequently, many of the options available to reduce GHG emissions involve improved energy efficiency. Current energy consumption is approximately 19 million British thermal units per ton of steel (MMBtu/ton) (22.1 GJ/tonne) for integrated mills and 5.0 MMBtu/ton (5.8 GJ/tonne) for EAFs. DOE estimates that a reduction of 5.1 MMBtu/ton (5.9 GJ/tonne) (27 percent) is possible for integrated mills (half from existing technologies and half from research and development [R&D]). A reduction of 2.7 MMBtu/ton (3.1 GJ/tonne) (53 percent) is possible for EAFs (two-thirds from existing technologies) (DOE, 2005).

ii. Nonferrous- Aluminum

Amongst the common non-ferrous metals, aluminum production by electrothermal reduction using carbon electrode, is a very high energy consuming process(175 GJ/ton compared to 20.5 GJ/ton for iron & steel)[12a] accompanied by large quantities of carbon di-oxide emission (12.7 tonnes/tonne of Al, 5 times that of steel, 14 times that of cement) (ref.12A). Additionally, the use of fluoride (cryolite) in the process generates PFC, such as CF_4 and C_2F_6, with CO_2 equivalent as 6500 and 9200 respectively[12a]. For a global production of 30 million tons of primary aluminium, about 390 million tons of CO_2 result in totaling 0.9% of total anthropogenic GHG emissions[12a]. Aluminum industry has been able to reduce emissions considerably due mainly to R&D efforts.

AlCOA (Aluminum Company of America) reached the goal of reducing GHGs emissions by 25% (from 1990 levels) in 2003, with increased aluminum production figures. The company believes that the aluminum industry can be "greenhouse gas neutral" by 2020. Alcoa has used hydroelectric power as a major energy source, for its operation around the world and evaluating the feasibility of building the world's first geothermal powered aluminum production plant.The electrochemical reduction process in a cryolite (fluoride) bath using graphite electrode, generates large volume of carbon dioxide gas and PFCs. The PFCs, such as CF_4 and C_2F_6, have very high global warming potential. Aluminum industry is working on 'inert anode' to replace graphite anode, which would stop producing carbon dioxide and PFCs. Worldwide use of 'inert electrode' to produce aluminum shall reduce GHG emission by nearly 40 million metric tons[12b]. In 2007, Alcoa launched "carbon capture" technology in their plant in Western Australia, In this process, carbon dioxide is mixed with bauxite residues (byproduct in the production process), thus locking up carbon dioxide. The resulting product is alkaline in nature, which can be used for road foundation, building materials, or an additive to improve soil.

e. Cement[1, 13]

Cement is considered as one of the most widely used material. Its production plant is one of the worst emitter of GHGs.The cement is made by heating limestone and clay until they fuse into a material called clinker, which is then ground and mixed with various additives to form cement. The heating by coal and the decomposition of limestone lead to generation of large quantites of carbon dioxide. Three main areas in which reduction in emission are planned, includes

- to make kiln more fuel efficient in order to reduce energy required per ton of clinker production.This is a key area of improvement.
- replace fossil fuel, like coal by farm wastes or used tyres. Environmentalists have started crying foul, since burning of waste release noxious chemicals.
- mix more additives into cement & thereby reducing proportion of emission intensive clinker. Building codes limit the extent the cement can be diluted
- waste heat recovery & use for power generation/ co-generation.
- overall energy efficiency improvement

The largest *opportunities for improving energy efficiency and reducing CO_2 emissions related to energy inputs* in the cement/ concrete industry are mostly related to cement manufacturing. The calcining of limestone ($CaCO_3$ = CaO + CO_2) produces 0.785 tonnes of CO_2 per tonne of CaO. There is no technological or physical way to reduce calcination reaction emissions. Cement contains about 61% CaO. Hence, the calcining reaction produces roughly 0.48 tonnes of CO_2 for each tonne of cement manufactured. Pyroprocessing accounts for 74% of the cement/concrete industries' energy consumption (93% of cement's manufacturing energy requirement) and operates at roughly 34% thermal efficiency.This low thermal efficiency provides many opportunities to improve performance. The improvement in energy efficiency has been achieved by switching over to 'dry' process from earlier 'wet' process, use of non-fossil fuels, cogeneration and other innovative practices. Realistic and cost-effective energy savings are smaller, but achievable[14].

It is reckoned by cement industries that about 5% of world's emissions of carbon dioxide are due to cement, which is double the emission figure of aviation industry. EU has imposed restriction on emissions from cement kilns. Also the major cement producers have an outfit called the Cement Sustainability Initiative, which imposed voluntary cut on emissions. Lafarge and Holcim pledges to cut emissions/ton of cement by a fifth by 2010 and Comex by a quarter by 2015. So far they are in track in their commitments.

Despite the above marginal measures for emission reduction per ton of cement, the increasing demand would result in higher total emissions. Cement industries have developed stronger and more flexible cements and concretes, the use of which by construction industries can lead to requirements of lesser quantities and thus lowering total emission figure. China is the largest producer of cement followed by India and thus the big emitters in this sector. Both the countries are on the right track to reduce emissions with the available technologies[15]. The United States ranks third in the world in overall cement output.

EU legislations on industrial emissions[16]

Carbon dioxide (CO_2) emissions from petrochemicals, ammonia and aluminium manufacturing industries will be included in the EU emissions trading scheme (EU ETS) post-2013, according to the draft climate change bill published by the European Commission (EC) on January, 2008. The bill must be approved by both the Council of the EU (national ministers) and the European parliament before it becomes law.The EC hopes that a final decision adopting modifications to the directive will be taken by 2009.

Summary

i. Nearly a third of the world's energy consumption and more than a third of carbon dioxide (CO_2) emissions are attributed to manufacturing industries.

ii. The large primary materials industries, *i.e.*, *chemical, petrochemicals, iron and steel, cement, paper and pulp, and other minerals and metals*, account for more than two-thirds of this amount.

iii. Of metals, aluminum production is a very high energy consuming process and also the worst emitter. Aluminum industry has been able to reduce emissions considerably due mainly to R&D efforts by aluminum major, Aluminum Company of America.

iv. Overall, industrial energy use has been growing strongly in recent decades. The rate of growth varies significantly between sub-sectors. For example, chemicals and petrochemicals, which are the heaviest industrial energy users, doubled their energy and feedstock demand between 1971 and 2004, whereas energy consumption for iron and steel has been relatively stable.

v. Despite the marginal measures for emission reduction per ton of cement by major cement producers, the increasing demand would result in higher total emissions.

vi. Steel industries in North America, Western Europe and Japan have reduced energy consumption per unit of production by 49 percent in the last 25 years. The steel industry now accounts for 3 – 4% of global man-made greenhouse gas emissions. Over 90 percent of steel industry emissions come from iron production in nine countries or regions: Brazil, China, EU-27, India, Japan, Korea, Russia, Ukraine and the USA.

vii. Paper and pulp industry in particular have reduced emissions by using innovative technologies.

viii. Trend is to produce more by using less energy.

REFERENCES

1. Executive Summary of the Report submitted by IEA for G-8 on studies to assess worldwide industrial energy efficiency in manufacturing sectors, 2005.http://www.iea.org/Textbase/npsum/tracking2007 SUM.pdf

2. New Energy and Industrial Technology Development Organization (NEDO), Japan:Global Warming Countermeasures' through "Japanese Technologies for Energy Savings/GHG Emissions Reduction, 2006 edition.

3. R.Chattopadhyay, *Surface Wear –Analysis, Treatment, and Prevention*, ASM International, Materials Park, OH, USA, 2001.

4. R.Chattopadhyay, *Advanced Thermally Assisted Surface Engineering Processes*, Kluwer Academic Publishing, MI, USA, (now taken over by Springer, USA), 2004.

5. Nitrous Oxide, *Source & Prevention,* EPA, Question & Answer http://epa.gov/nitrousoxide/sources.html

6. Marta Khrushch, and others, *Carbon Emissions Reduction Potential in the US Chemicals and Pulp and Paper Industries by Applying CHP Technologies*, Proceedings of the 1999 ACEEE Summer Study on Energy Efficiency in Industry, Saratoga Springs, NY, June 15-16, 1999

7. Combined Heat and Power in the Indian Pulp and Paper Sector - Thomas D. Schmitz, handed to Journal:"Industrial Cogeneration"on 31.03.2008. http://www.energymanagertraining.com/ CHP/download/CHPintheIndianPulpandPaper.pdf

8. Available and emerging technologies for reducing greenhouse gas from the iron and steel industry, EPA, United States Environmental Protection Agencies, October,2010;http://ww w.epa.gov/nsr/ghgdocs/ironsteel.pdf

9. Iron & Steel Emissions. Globalgreenhouse.com;http://www.global-greenhouse -warming.com/ iron-and-steel-emissions.html

10. International Energy Agency, OECD, 2007, website on 'Tracking Industrial Energy Efficiency and CO2 Emissions.

11. Reuters reports in Varian, (Berlin, Tuesday, Oct8, 2007

12a. P.K. Sen, guest editorial, IIM Metal News, vol14,no,3,3June,2011

12b. ALCOA 's green R&D, Span, april/march, 2008, p. 17.

13. Cement Plant Emissions, *The Economist*, and 22nd Dec, 2007.

14. William T. Choate; Energy and Emission Reduction Opportunities for the Cement Industry, December 29, 2003, Energy Efficiency and Renewable Energy, US Department of Energy

15. *GHG Reduction Options - Indian Cement Industry;* Jyoti Parikh,. B.D. Sharma, Vineet Kumar,&. Arshi Vimal; Conf. on Options to Reduce GHG Emissions through National Appropriate Mitigation. 13 May 2009, Beijing

16. EC Takes Steps to Cut Industrial Emissions; engineers.ihs.com/news/eu-en-industrial-emissions-12-07.htm?WBCMODE=presen\\\\\\\\\\\\%255

15

Transportation

The movement of men and materials by road-, rail-, air- and sea routes are carried out by automobiles, trains, planes and ships respectively. The industries responsible for these activities are known as transportation industries. All these industries use fossil fuels for running the engines. Large quantities of emissions not only pollute the atmosphere but also increase the GHGs concentrations. The various measures adopted for reducing emission include, more efficient running of engine to get more mileage per unit of fuel used, improving combustion efficiency, less energy loss in the transmission chain, and partial replacement by alternative fuel. This chapter shall deal with these measures to reduce GHGs emissions in different sectors of the transportation industries.

Emission by Transport Sector

The figures for transport of carbon dioxide in EU have increased by about 33% in between 1995 and 2005, thus constituting 27% of total emission[1]. Of the 27% of total carbon dioxide emission in transport sector, emissions by different transport modes, in 2005, are given in table 14.1

Table 14.1. Carbon dioxide emission by various transport modes

Transport mode	*Carbon dioxide emission (million tons)*	*Percentage*
Road Transport	696	72%
Sea & Inland waterways	189	15%
Aviation	150	12%
Railway	20	1.6%
Others	10	0.8%

In USA, transportation sector carbon dioxide emissions in 2008 were 95.6 MMT lower than in 2007 but still 343.2 MMT higher than in 1990. The transportation sector has led all U.S. end-use sectors in emissions of carbon dioxide since 1999; however, with higher fuel prices and slower economic growth in 2008, emissions from the transportation sector fell by 4.7 percent from their 2007 level. Petroleum combustion is the largest source of carbon dioxide emissions in the transportation sector. Increases in ethanol fuel consumption in recent years have mitigated the growth to certain extent in transportation sector emissions. Reported emissions from energy inputs to ethanol production plants are counted in the industrial sector emission[2]. In India, during 1997-2004, transport sector emissions grew 4.5 per cent to 142 million tonnes[3].

Emission Management in Transport Industries

The steps adopted to manage emission include the followings:

- more efficient running of engine
- improving combustion efficiency
- less energy loss in the transmission chain, and
- full or partial replacement by alternative fuel

In transport systems using fossil fuel the emission reduction is assessed by achieving more mileage per unit of fuel used.

Road Transport –Automobile

Calculation of CO_2 Emission by Vehicles[1,4,5]

Unleaded gasoline has 8.87 kg (19.56 lbs) of CO_2 per gallon. By dividing number of miles driven by miles per gallon, the figure for number of gallons of gasoline consumed annually is arrived. The multiplication of this figure by 8.87 and division by 1,000 would lead to metric tons of CO_2 emitted[2]. The road transport[4,5] has the largest share of carbon dioxide emission (72%) in the transport sector.This translates to 19.44% or one fifth of global emissions. Automobiles—by which we mean personal motor vehicles, including light trucks such as pickups, SUVs and vans as well as sedans and wagons—emit roughly 10% to 13.5% of global CO_2 emissions. Rest 6 to 9.44% emissions are due to heavy vehicles.

Steps to the reduction in emission loss in automobile sector

In spite of emerging alternative propulsion technologies the internal combustion (IC) engine is likely to have a place in the automotive sector for the next 50 years, likely longer in the developing world. Improving the efficiency of the IC engine can therefore play an important role in reducing global CO_2 emissions and provide customers with greater fuel economy. Some major steps to improve fuel efficiency include followings:

1. **Reduction in car weight:** The yield stress of high strength microalloyed steel is around 33% higher than low carbon rimming steels. Therefore the convention rimming steel strips are replaced by thinner high strength microalloyed steels in making the car body. The reduction in dead load by 30% leads to correspondingly less fuel consumption for the same payload. The replacements of cast iron engine components by lighter aluminum alloys and metallic mud guard, dashboard by plastics etc have led to further reduction in weight.
2. **Tribo-science-to negate part of 40% energy loss in transmission:** Triboscience is the study of friction, wear and lubrication. Friction or resistance to motion by mating components (*e.g.* gears, cams) results in energy loss in transmission[6,6a]. Researchers in triboscience (study of friction, wear & lubrication) at EU Tribo Research Commission's funded research have found that about 40% of the available energy in an automobile is lost due to

friction and wear in the transmission, plus energy lost in cooling and exhaust, leaving only a minor portion of the energy transmitted to wheels. Researchers are finding ways and means to minimize these frictional losses in order to reduce carbon dioxide emission to the level committed by ACEA[7]. Surface engineering and surface lubrication are used to minimize wear and frictional losses.

3. **Improving Engine Efficiency:** Some of the measures adopted to improve engine efficiency include, improvement in combustion efficiency, minimizing heat lost by using insulating coating on piston crown & valve face, erosion resistant coating on valve seat etc. The insulating zirconia coating on piston crown can reduce 2% diesel consumption in heavy vehicles[6a].
4. **Use of alternate fuels:** Current practice is mostly to use 10% biofuel (bioethanol) mixed with petrol. Hybrid cars have been developed and introduced in the market.

EU and Emission limits

The European Association of Car Manufacturer (ACEA) has committed to reducing emission levels to 140g/Km in 2008 and to 120g/Km in 2010-2012. The current emission rate in the car is 160gm CO_2/Km (g/Km). Emission was reduced at a rate of 1.5% instead of required rate of 3% to achieve the voluntary target of 140g/Km set by automobile industry. EU have a new norm for new cars to be on road in EU countries by 2012. The new proposed limit is 130g/Km. Another 10g less than present norm, expected to get from other sources than the existing practice, such as, low rolling resistance tyres, more efficient A/C, and greater use of biofuel [1].

Automobile industries' responses to EU's new norm, mostly positive barring few, as follows[1]:

i. **French, Italian, such as,PSA Peugeot, Citroen, Renault, and Fiat:** Their 2006 fleets are mostly fuel-efficient, small cars with average emission figure of 142-147g/Km. In order to conform to new norm, they may need to increase the price of the cheap, low margin cars, which they are selling now.

ii. **Germans:** Small cars made by Volkswagen, including Blue Motion Polo have already achieved a low emission figure of 99g/Km, less than 104g/Km of Toyota Prius hybrid. However the *fleet average emissions* of these cars have increased due to Audi brand.

Premium brand includes Audi, Mercedes-Benz, BMW are the one finding it difficult to cope with the EU's norm. Mercedes in 2006, is in the top of emission league with fleet average of 188g/Km, followed by BMW with fleet average figure of 184g/Km. BMW achieved a figure of 128g/Km for their new 3-series cars of 2-litre, diesel, 177 brake horse power(bhp) by (i) reducing car weight and (ii) adopting new technologies, such as, in alternators and coolant pumps, automatic start-stop to shut engine in stationary traffic, and regenerative brake. But for heavier premium brand cars, like Mercedes being the worst emitters, the commission would like to fine on $137(95 euro) per car per gram of emissions exceeding 130g/Km. However commission has also agreed to "weight dispensation" *i.e.*, charging heavy cars according to slope of the weight/ CO_2 graph. Without weight concession Mercedes fleet would attract a penalty of about 5,500 euro per car.

Other countries are also tightening control of automobile emissions. China has imposed fuel economy regulations. American Congress has approved a bill tightening control and in California, all petrol sales shall require a low-carbon standard.

USA[1,8]

U.S. automobiles and light trucks are responsible for nearly half of all greenhouse gases emitted by automobiles globally. The Big three automakers—General Motors, Ford and Daimler Chrysler—accounted for nearly three-quarters of the carbon dioxide released by cars and pickup trucks on U.S. roads in 2004[8]. Cars and trucks made by GM gave off 99 million metric tons of carbon dioxide or 31 percent of the total. Ford vehicles emitted 80 million metric tons or 25 percent. Daimler Chrysler vehicles emitted 51 million metric tons or 16 percent[8]. While Americans own only 30 percent of the 700 million vehicles that are in use worldwide, the authors of the report found that cars in the U.S. account for a disproportionate amount of greenhouse gas emissions because they are driven farther, have lower fuel economy standards, and burn fuel with higher levels of carbon than many of the cars in other countries.

Zero Pollution Motors—CAV Car[9]

A compressed air car, or CAV technology is based on compressed air to drive old-fashioned car engine pistons. can be used instead of combusting gas or diesel fuel. The air car can run at a top speed of 35 mph for some 60 miles on a tank of compressed air, a sufficient distance for 80% of urban commuters to complete daily chores. On highways, the CAV can cruise at interstate speeds for nearly 800 miles with a small motor that compresses outside air to keep the tank filled. The motor can use any fuel, such as, gasoline or diesel as well as biodiesel, ethanol or vegetable oil. This car leads the highest-mpg vehicles available right now. Even if it used only regular gasoline, the air car would average 106 mpg, more than double of today's hybrid, the Toyota Prius. The air tank also can be refilled when it's not in use by being plugged into a wall socket and recharged with electricity as the motor compresses air.

Indian car maker Tatas have already bought the rights to make the car for the huge Indian market. Tata doesn't plan to produce the cars in the U.S. Instead, it plans to charge $15 million for the rights to produce and market the same.

Aircraft-aviation Industry

Air Travel – CO_2 emissions

The carbon emissions in air travel vary by length of flight. Emission figures amount to 0.24 kg CO_2 per passenger mile for short flights down to 0.18 kg CO_2 per passenger mile for long flights[8]. Boeing expects the number of commercial jetliners to nearly double, to 36,420, in the next 20 years. The Federal Aviation Administration expects 1.2 billion passengers a year to travel on

U.S. carriers by 2020, up from 741 million last year. By 2050, the industry is expected to contribute anywhere from 6 to 10 percent of the gases and particles tied to global warming, up from about 3 percent today[10].

European emissions-trading proposal would require all airlines flying through domestic routes to enter into an emissions-trading scheme by 2011. Carriers flying to and from Europe, including U.S. airlines would have to enter the system by the following year. The plan is based on one already in operation for other European industries that buy and sell credits to emit certain amounts of carbon dioxide. The United States and other nations plan to vigorously fight the proposal[10].

Airplane and engine makers have set a goal of reducing carbon emissions on the next generation of jets by 50 percent by 2020.

EU looks at greener air travel. A 1.6 billion Euro research programme that will investigate products and manufacturing processes to make air travel more sustainable has been launched by the European Commission. The seven year 'Clean Sky' project will involve 86 organizations from 16 countries, including, research centres and universities. Its aim is to include 50% reduction in CO_2 emissions by reducing fuel consumption, an 80% reduction in nitrous oxide emissions, halving external noise, and a green design, manufacturing, maintenance and improvement in product life cycle. The use of advanced materials leads to lighter aircraft, better aerodynamics, improved systems and advanced engines result in lower fuel burn and less noise[11].

The recent summit on aviation and environment[12] has stressed the need to use environmental technologies, including the use of biofuel and advanced traffic management systems to reduce carbon emission. This would result in achieving 25% fuel efficiency target and more importantly shall lead towards the vision of a carbon emission free industry[12]. Taking into account that the aviation sector represents only 2% of total man-made CO_2 emissions (responsible for climate change), reducing air traffic is not a solution to the sustainable development of our world. Indeed, reducing air traffic by 50% would result in decreased CO_2 emissions of 1% while negatively impacting GDP by 4%. Flying shortest distances between possible destinations, reducing congestion around airports and increasing efficient use of aircraft are areas that the industry is working on jointly.

Asset management

Materials prognosis and the asset management practices have led to vast improvement in the useful life of the aircraft beyond the designed life in US military (fig 14).This step has improved reliability of the components combined with enormous financial, materials and energy gains, which would otherwise have been spent in acquiring new aircraft[13]. The prognostics and asset management have resulted in prolonging the life of F-16 aircraft to 33 years from the design life of less than 10 years (Fig.14)[13]

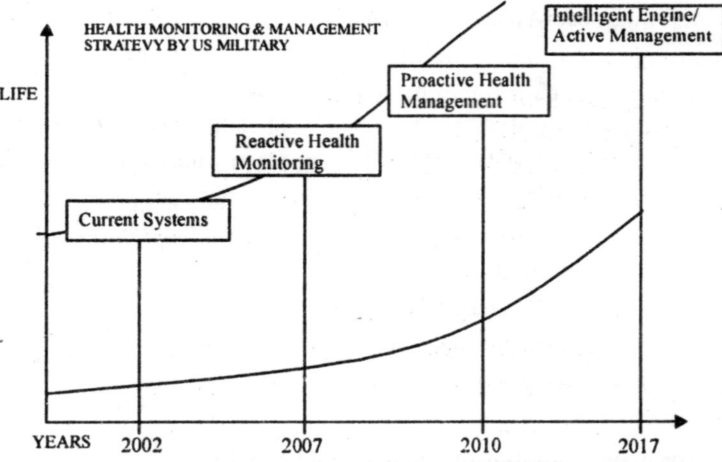

Fig. 14.0 Life cycle extension of military aircrafts by material prognosis & asset management (13)

Increase in combustion temperature in gas turbines

Another way of improving fuel efficiency is to increase the combustion temperature. This has been possible by improving the thermal properties of the metallic engine materials by ceramic thermal barrier coatings which can withstand high combustion temperature[14].

Maintain, Repair & Operate (MRO) aircraft

Repair and maintenance shops for the airlines are responsible for improving the life of the engine and reliability of operation. The regular maintenance schedules for overhauling, repair and refurbishing the engines and particularly the recoating of worn thermal barrier layer have led to improvement in operating energy efficiency in aero-engine

Airlines Green Program

Some airlines have initiated green program of afforestation in their carbon offset projects to reduce emission. For example, Northwest Airlines, apart from donating $1 million for land conservation projects around their domestic hubs and a carbon offset project in the Lower Mississippi River Valley.Other programs of Northwest Airlines include: (i) replacing old DC-10 by more fuel efficient A330 aircrafts, in trans-Atlantic route has resulted in 35 % more fuel efficiency compared to earlier aircrafts (ii)in 2007, NWA used 575 million fewer gallons of fuel than in 2000, the equivalent of more than 1 million passenger cars off of the road for a year and (iii)by introducing Boeing 787 Dreamliner, expected to save 20% more fuel than today's aircraft of comparable size.[15]

Rail Transport

Rail Travel - Carbon dioxide Emissions

The CO_2 emissions for rail travel vary by distance of the trip. On average, commuter rail and subway trains emit 0.35 lbs CO_2 per passenger mile, and long distance trains emit 0.42 lbs CO_2 per passenger mile[16]. Transportation conditions vary in real life beyond that can these estimated figures.To ensure that the rail calculators fully covers trip, 10% is added to the total mileage of the trip.

European Railway and Infrastructure Companies (CER) and the International Union of Railways (UIC) have made in-depth analysis of emissions in rail & other transport[17]. Some important points are as follows:

i. Traveling by rail is on average 3-10 times less CO_2 intensive compared to road or air transport. In comparison to other transport modes, Rail is the only one that has decreased its share of CO_2 emissions since 1990.

ii. With 7-10% of market share, rail still contributes less than 2% of the EU transport sector's CO_2 emissions. The rail sector has committed itself to cut the specific emissions of rail transport by 30% over the period 1990-2020.

iii. About 80% of the European rail fleet runs on electric power, meaning most trains can switch to clean electricity when it becomes available.

Use of biofuel

Virgin Trains plans to run its fleet of diesel-powered Voyager trains, which operate on the Cross Country Penzance-to-Aberdeen franchise, on a blend of biofuel and diesel as part of an environmentally friendly scheme. Biodiesel, made from plants can be carbon-neutral if the carbon dioxide emitted in burning can get absorbed by growing plants from the atmosphere. However, the higher duty on blended biofuel and diesel has been a concern in using the same as alternative fuel. The Virgin fleet uses around 90m litres of diesel per year and generates 3.2kg of carbon per mile[18].

Indian Railways[19,19a]

The rail network traverses the length and breadth of the country, covering a total length of 64,015 kilometres (39,777 mile). It is the 4th largest railway network in the world, transporting over over 1050 million tonnes of freight annually. The most valuable asset is the huge railroad network, which is to be maintained over 24x7 periods.The increase in the axle load and speed over the decades have led to higher rates of wear of the rail road, specially the rail crossings or frogs, where the wheels change the track. The development of advanced maintenance and repair systems (M&R) have been responsible for prolonging the life of frogs (crossing) enabling them to carry higher GMT (gross million tons) at high speed over a longer period without replacement of the crossings[19].

The Indian Railways have used mixtures of diesel plus various proportions of the oil from the Jatropha plant to power its diesel engines.Currently the diesel locomotives that run from Thanjavur to

Nagore section,Tiruchirapalli to Lalgudi, Dindigul and Karur sections use a blend of Jatropha and diesel oil. India Railway have plans to implement a Siemens' project involving use of special types of locomotives that utilise the momentum of trains running on electricity to regenerate the energy dissipated during braking. This energy can be utilized by another train in the section, running on the same overhead electric cables. The process results in 10-30 per cent savings in electricity consumption for the railways[19a].

Ships and Marine pollution

In comparison to road and air, sea transport is an energy efficient means of moving huge volumes of freight across the globe. Normally low grade, heavy, poor quality oil powers much of the deep sea shipping fleets. Poor fuel burnt in marine diesel engines exhausts nasty emissions into the atmosphere including particulate matter, carbon oxides and volatile organic compounds. Experts estimate that sea going ships exhausts now emit up to 2.8% of the world's greenhouse gas emissions[20]. New European Union legislation will require strict emission monitoring standards for ships in Port from 2010. Lloyds Register of Shipping has come up with a model for the accurate measurement of ship exhausts. A company in England has produced a ship's exhaust scrubber which it claims removes up to 95% of the nastiest from ship exhausts.

A solution being developed by Mitsubishi Heavy Industries, Japan, includes fuel_water injection system for ships diesel engines and sea water electrolysis and scrubber units. A new generation of ocean going ships now on the drawing boards will burn a variety of heavy residual fuels with zero or next to zero emissions, as stated by MAN Diesel Company.

Summary

1. The figures for transport industries' carbon dioxide emissions in EU have increased by 33% in between 1995 & 2005, thus constituting 27% of total emission. Of the 27% of total carbon dioxide emission in transport sector in 2005, road transport accounts for 72%, sea & inland waterways for 15%, aviation industry for 12%, railway for 1.6%.

2. Road transport industry accounts for almost 3/4th of the emission in transport sector. Stringent emission limits by EU & others have led the automobile manufacturers to make more fuel efficient cars. Other effective measures include, use of biofuel mixed with petrol, and hybrid cars.
3. Aviation industry has taken effective steps to improve fuel efficiency of the engine. Proper asset management and MRO shops have led to improvement in engine life and reliability thus enabling further improvement in the emission figure. Airlines also have taken effective steps to reduce fuel consumption and to reduce carbon offset through participation in green projects.

REFERENCES

1. *The Economist*, December, Lower Emission Cap & Automobile Industry 22nd, 2007)
2. Emissions of Greenhouse Gases Report, DOE/EIA-0573(2008), US Energy Information Administration
3. *Business Line*, business daily from The Hindu,May12, 2010
4. Carbon Calculation in Transportation http://www.carbonfund.org/site/pages/carbon_calculators/category/Assumptions/#Transportation#Transportation
5. Summary of Travel Trends 2001, National Household Travel Survey, Pat S. Hu and Timothy R. Reuscher December 2004,U. S. Department of Transportation, Federal Highway Administration, http://nhts.ornl.gov/ 2001/pub/STT.pdf
6. R.Chattopadhyay, *Surface Wear-Analysis, Treatment and Prevention*, ASM International,Materials Park, OH,USA,20001

6a. R.Chattopadhyay, *Advanced Thermally Assisted Surface Engineering Processes*, Kluwer Academic Publishers (now taken over by Springer), MA,USA, 2004

7. European Concerted Action on "Triboscience and Tribotechnology-: Superior friction and wear control in engines and transmissions" COST 532; Coordinator: Dr. Amaya Igartua,Spain.EU triboresearch commission, tds.ec-lyon.fr/cost516/.../C532-PP-M1-01-02.pdf

8. Global Warming on the Road- the climate impact of America's automobile, John DeCicco and, Freda Fung, of Environmental Defence, http// www.edf. org/ document/5301_Globalwarmingonthe road.pdf.

9. An article by Jim Ostroff, Thursday, October 30, 2008, New York Times.

10. U.S. Airlines Under Pressure To Fly Greener, *By Del Quentin Wilber.* Washington Post Staff Writer, Saturday, July 28, 2007; Page D01.

11. Sources: Emissions factor based on 2006 data collected in the UK available from the The World Resources Institute.

12. 'Aviation and Environment' summit at Geneva (22-23 April, 2008) the chairman & CEO,IATA, Mr. Giovanni Bisignani. The Statesman, Kolkata edition, 26 May, 2008, p. 9.

13. J.M. Larsen & others: *Achieving the Potential of Materials Prognosis for Turbine Engines*, DARPA Bidders Conference on Materials Prognosis, 26th September, 2002

14. *Materials World*, March, 2008,p4

15. North-West airlines website

16. Source: WRI Employee commuting spreadsheet

17. European Railway and Infrastructure Companies (CER) and the International Union of Railways (UIC) in their publication on 'Rail Transport &Environment–Facts&Figures', in June 2008 (http:// www.uic.asso.fr/ homepage / railways& environment facts &figures .PDF.

18. Dan Milmo, *The Guardian*, Monday October 16 2006.

19. R.Chattopadhyay, Rail Wheel Contact Wear,Intl. Symp. On Tribology, 18-23 Feb,1993, Beijing,China, National Science Foundation of China,Chinese Mechanical Engineering Soc., and Tsinghua University.

19a. Business Line, *Financial daily from The Hindu*, India, October 20, 2005.

20. Ships to Embrace Energy Efficiecy Measures:Environment News Network, April7, 2010, http://blog.cleantechies.com/2010/04/07/ships-energy-efficiency-bunker-fuel.

16

Conservation of Biomes and Buffers

Biomes are the habitants of a large number of plants and animals, in lands and oceans, under different natural climatic conditions. Their existences directly and indirectly play a significant role in maintaining natural cycles and climates. Major pools of carbon include the atmosphere, fossil fuels, oceans and the terrestrial biota (animals & plants) and soils. In natural cycle, carbon is exchanged between these pools and the atmosphere. Ideally net carbon gain by the atmosphere due to the exchange processes should be zero. The atmospheric concentrations appear to have increased rapidly over the past 100 years and are currently higher than ever in human history. This suggests that more carbon is being released to atmosphere than can be absorbed on pools. A major cause for more carbon in the atmosphere is the decreasing capacity of the pools acting as sinks, such as, forest, ocean and terrestrial biota. Over the past several decades, increasing human activity has caused rapid destruction of forests, pollution of oceans, and elimination of biota, with consequent shrinkage of capacity to absorb carbon. To reverse this trend it is important to put concerted efforts to prevent further destruction of the biomes and also to find ways and means for revival and /or increasing in the capacities of these biomes to absorb carbon.

Global Carbon Cycle-forests and Oceans

A reservoir with the capacity to store and release carbon, such as soil, terrestrial biosphere, the ocean, and the atmosphere is called *carbon pool*. The ocean contains the largest active pool of carbon near the surface of the Earth, but the deep ocean part of this pool does not rapidly exchange with the atmosphere. The global carbon cycle involves the natural emission, absorption and storage of huge quantities of carbon. It has been estimated that, every year, nearly four giga tonnes of carbon are exchanged between the earth / oceans and the atmosphere (table 15.1). The change in carbon concentrations due to emission and absorption is known as the *carbon flux*. Net emitters of carbon are known as carbon sources and net absorbers of carbon are known as carbon sinks. Carbon can be stored in a number of ways such as in coal or oil, oceans, organic matter and plants. Carbon is added to the atmosphere through the burning of fossil fuels, organic respiration, wood burning, and volcanic eruptions. The uptake of carbon from the atmosphere occurs through mainly carbon dissolution into the oceans, and plant respiration.

Since the 1940s, tropical deforestation has accounted for by far the greatest net emissions from the natural environment, although these are still far below emissions due to fossil fuel use. In 1990 fossil fuels accounted for the release of some 6 billion tons of carbon into the atmosphere whereas deforestation accounted for approximately 1.7 billion tons.

Global Carbon Emission and Flow

The total emission of 7.7Gt in the atmosphere is reduced to 3.7 Gt through absorption of carbon by oceans and forests of 2 billion tons each (table 15.1). Prevention of deforestation could have changed the emission to atmosphere figure to a low value of 0.3 billion tons (almost carbon neutral), due to increase in the absorption capacity of forests by 1.7 billion tons[1a,2].

The greatest stores of carbon are believed to be the world's oceans and fossil fuel reserves. On land, carbon is stored in ground litter, soils and plants.

Table 15.1: Approximate current net annual global carbon flux as Gt (billions of tons) of carbon

Activity	*Source+/Sink-(Gt)*
Absorption by Forests	-2
Absorption by Oceans	-2
Deforestation	+1.7
Fossil Fuel Use	+6
Total	+3.7

In Fig.15.0, the quantities of carbon storage and annual carbon exchanges on the earth's surface are shown. According to fourth IPCC report (1b), estimated amount of carbon stored in the atmosphere is 760 PgC (Pg= 10^{15}g, 10^{12}kg, 10^9t, 1Gt) of CO_2 and its amount is increasing at rate of 3 PgC/yr. On the other hand, the amount of organic matter stored in terrestrial ecosystem is 2000 PgC, which is about 3 times of that in the atmosphere. In addition, about 2/3 of the organic matter is contained in the forest ecosystem. Therefore the variation of the organic matter content in the forest ecosystem would have significant impact on atmospheric CO_2 concentration. The IPCC report estimated that the forest ecosystem is neither sink nor source of CO_2, in other words, the input and output of carbon into and out of the system compete each other. However, these estimations are not accurate enough, so that reliable estimation of the carbon budget derived from the results of measurements in various ecosystems which cover wide area is urgently required.

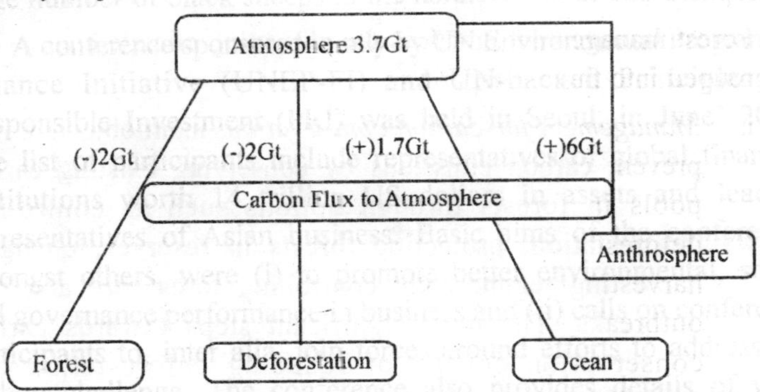

Fig. 15.1 Global scale Carbon Flux and from Atmosphere

Forest carbon "flux" can be more precisely measured by 'Eddy Correlation Flux Measurement'. The system provides *in situ* half hour averages of the surface vertical fluxes of momentum, sensible heat, and latent heat. The fluxes are obtained by the eddy-correlation technique, *i.e.* by correlating the vertical wind component with the horizontal wind component, the sonic temperature (which is approximately equal to the virtual temperature), and the water vapor density. A 3-dimensional sonic anemometer is used to obtain the orthogonal wind components and the sonic temperature. An infrared hygrometer is used to obtain the water vapor density[3].

Forest Management

Carbon management of forests requires data on carbon pools and carbon exchange or flux with atmosphere. The theoretical models[1,2] and simulation studies are used in profiling carbon pools and fluxes. Forests are thought to contain about 80% above-ground and 40% below-ground of total terrestrial organic carbon. The world's forest pools contain more than 55 percent of the global carbon stored in vegetation and more than 45 percent of that in soil. Most of the carbon pool in forest vegetation is located in tropical forests (62 percent), whereas most of the carbon pool in forest soils is located in boreal forests (54 percent)[2]. Carbon flux data in tab.15.1 indicates the need to prevent deforestation and to utilize deforested land for planned reforestation. Reforestation can absorb a substantial portion of the 1.7Gt of carbon flux, earlier lost to atmosphere due to deforestation.

Forest management practices for mitigating climate change can be grouped into three categories[2]:

i. *Management for carbon conservation:* main objective is to prevent carbon emissions by conserving existing carbon pools in forests through options such as controlling deforestation, protecting forests in reserves, changing harvesting regimes and controlling forest fire and pest outbreaks. The most significant steps towards carbon conservation consist of reducing deforestation and degradation in the tropics. In recent years, there has been significant expansion of "protected areas", comprising both mature forests and forests in other development stages, for the conservation of biological diversity and watershed protection. For example, Costa Rica protects 25% of its national territory within the *protected area system,* which is the largest percentage of protected areas in the world. It also possesses the greatest density of species in the world. While the country has only about 0.1% of the world's landmass, it contains 5% of the world's biodiversity[4]

ii. *Carbon storage management*: goal of storage management is to increase the storage of carbon in the vegetation and soil of forest ecosystems by increasing the area and/or carbon density of natural and plantation forests, and also to increase its storage in durable wood products. Increasing the carbon pool in vegetation and soil can be accomplished by protecting secondary forests and other degraded forests whose biomass and soil carbon densities are less than their maximum value, thereby allowing them to sequester carbon by natural or artificial regeneration and soil enrichment. Sequestering carbon by storage management is only a short-term option, producing a finite carbon sequestration potential beyond which little additional carbon can be accumulated

iii. *Ccarbon substitution*: aims at the substitution of forest biomass carbon for non-renewable sources of raw material and fossil fuel-based energy, such as construction materials and biofuels. This approach involves increasing the use of forests for wood products and fuels, obtained either from establishing new forests or plantations, or increasing the growth and subsequent potential fibre production of existing forests through silvicultural treatments. Silviculture (Latin 'silvi' means forest) is the practice of controlling the establishment, growth, composition, health, and quality of forest to meet diverse needs and values. Over long periods, the substitution management method is likely to be more effective in reducing carbon emissions than the physical storage of carbon in forests or forest products[5].

The implementation of forest management options that are compatible with traditional objectives of forestry, over the next 50 years or so there is a potential to conserve and sequester an amount of carbon equivalent to about 11 to 15 percent of total fossil fuel emissions over the same period. The adoption of forest management options that conserve and sequester carbon would help to prevent forests from becoming a significant net source of CO_2 for the atmosphere in the future and thereby help to offset other factors that contribute to accelerate global warming[2].

Trees/Forests and the Global Carbon Cycle

The photosynthesis converts carbon dioxide into carbohydrate, which is stored as biomass in the tree, and oxygen is released back into the atmosphere (Fig.15.2). Young trees grow more rapidly and absorb more carbon dioxide than old trees. Although the old trees absorb less carbon dioxide, they have much greater stores of carbon in their biomass. In Scandinavia, for example, trees can live for up to 700 years, storing carbon for long periods. However, they eventually die and rot releasing the stored carbon back into the atmosphere.

Carbon absorbed by trees also returns to atmosphere, when the products, such as timbers and paper are incinerated, or when they rot in landfill and release methane - another, more potent, greenhouse gas. In order to alleviate climate change there must either be an increase in the amount of carbon that is taken from the atmosphere and stored for long periods or there must be a reduction in anthropogenic carbon emissions (or a combination of both). It has been suggested that forests and forestry have the potential to achieve these through an increase effort in the aforestation.

Forest Preservation, Kyoto and Bali Conference

The Kyoto Protocol recognizes that forests play a key role in global warming, since they are both sources and sinks of carbon dioxide. In fact, forest loss to agriculture or development, along with over harvesting, have made forests the second largest source of CO_2. However, as Article 2 of the Protocol states, when existing forests are conserved and sustainably managed, or cut-over forests are replanted, they become effective long-term sinks.

When Kyoto Protocol was negotiated, deforestation was excluded partly because of the absence of a reliable scientific method for determination of 'carbon loss' due to deforestation. Hence under the current regime of Kyoto Protocol, which expires in 2012, *financial rewards for storing carbon through trees were attached only to reforestation or planting new trees*. However, this restriction became hard to justify as evidence mounted that the destruction of tropical forests caused as much as 20% of GHG emissions.

With the availability of better technology for assessment in carbon loss, Bali conference's call for 'meaningful action to reduce emissions from deforestation and forest degradation' may pave the way to forest preservation. *Brazil and Indonesia ranks the world's third- and fourth- largest emitters, if deforestation is taken into account* (see Chapter 9).

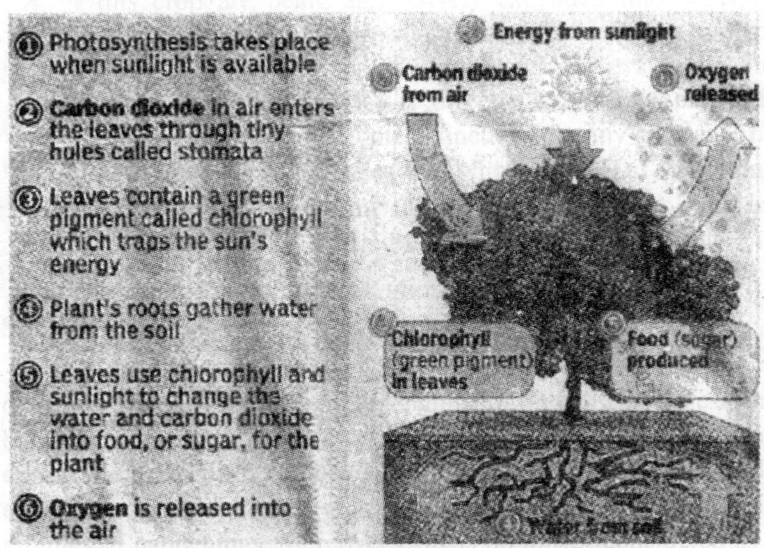

Fig. 15.2 Mechanism of Removing Carbon by Trees

Deforestation

Deforestation is the process of conversion of forest to another land use or the long-term reduction of the tree canopy cover below a 10 percent threshold. Deforestation implies the long-term or permanent loss of forest cover and its transformation into another land use. Forests are cleared, degraded and fragmented by timber harvest, conversion to agriculture, road-building, human-caused fire, and in myriad other ways. The forests have been mostly removed for fuel, building materials and to clear land for farming. About one half of the forests that covered the Earth are gone. Each year, another 16 million hectares disappear.

The World Resources Institute estimates that only about 22% of the world's (old growth) original forest cover remains "intact" - most of this is in three large areas: the Canadian and Alaskan boreal forest, the boreal forest of Russia, and the tropical forest of the northwestern Amazon Basin and the Guyana Shield (Guyana, Suriname, Venezuela, Columbia, etc.). However the coverage of these forests are low compared to tropical forests. Slightly more than 50% of the forests are found in the tropics and the rest are temperate and boreal (coniferous northern forest) zones.

Today, forests cover more than one quarter of the world's total land area, excluding Polar Regions. Seven countries (Russia, Brazil, Canada, the United States, China, Indonesia, and the Democratic Republic of Congo (formerly Zaire) account for more than 60% of the total.

Fig. 15.2 Roadbuilding and Deforestation (picture) by authors, Bhutan forest road to India, 04-06-2009

In the last few decades, the vast majority of deforestation has occurred in the tropics - and the pace still accelerates. The removal of tropical forests in Latin America is proceeding at a pace of about 2% per year. In Africa, the pace is about 0.8% per year and in Asia it is 2% per year.

Deforestation Processes

The natural forests of the world can be split broadly into two main categories. The tropical forests lay either side of the equator and the temperate and boreal forests are largely found in the northern hemisphere (with a small proportion in Australia, New Zealand, Argentina and Chile).

Tropical forests in Latin America, India and other South East Asian countries have fallen victim to:

- timber exploitation/logging
- slash and burn farming,
- clearfelling for industrial use
- road building
- cattle ranching
- increasing demand for meat, skin, tusk etc are causing rapid decrease in animals and other species.

Deforestation in the tropics is the second most important source of greenhouse gas emissions *Over half of the world's original tropical forests are already gone*. Tiger population in India has gone down to the lowest level. Public attention to this exploitation by media & NGOs have helped to alleviate the problem somewhat, though many challenges are still to be faced. In Indonesia, the world's third largest emitter (mainly due to deforestation) 80% of deforestation is due to illegal logging which serves mainly the pulp and timber industry[6].

Boreal forests are subjected to deforestation due to similar reasons as that of tropical forests. Boreal forests are found in the broad belt of Eurasia and North America: two-thirds in Siberia with the rest in Scandinavia, Alaska, and Canada. Logging old-growth forests and replacing them with plantations intended for timber/ paper production results in a net loss of carbon which is released into the atmosphere. This is especially relevant in Canada, Russia and the Baltic States where this is most widespread, and also in lesser extent in Scandinavia

Road building, mining, and host of other big industries, such as, integrated steel plant, thermal and hydroelectric power (construction of big dam) plants, cement etc, have led to deforestation of enormous areas in the developing nations with higher growth rate. For example, road building in pristine forest area has resulted in deforestation (fig15.2) for hundreds of miles along the tracks in the mountain range connecting Bhutan to India. The repair, maintenance and road-widening processes are in progress throughout the year, thus causing further deforestation. Human habitation across the transport routes has resulted in deforestation in the adjacent areas. This being only transport route between India and Bhutan, thousands of vehicles pass through this route day and night with man and materials, polluting the atmosphere. In addition to this, construction of hydroelectric plants in the downstream rivers has led to further deforestation of vast areas. Deforestation, industrial activity and the use of fossil fuels have elevated carbon dioxide levels in Earth's atmosphere by 25 percent over the last 100 years. Concentrations of this heat-trapping gas continue to rise. *The most important forest measure to curb climate change is to halt deforestation.*

Afforestation and forest industries

Increasing the area of forests can only be a temporary measure as forests eventually stop absorbing CO_2, unless the rate of afforestation becomes more than that of carbon removal. Afforestation could provide a "window of opportunity" in which to cut carbon emissions, but it is important that it is carried out in an ecologically and socially sustainable manner. Current rates of forestation absorb enough carbon to offset just 7-16% of the annual increase in carbon emissions let alone reduce overall levels. In an UN Environment Programme (UNEP), for planting one billion trees in 2007, the total figure has exceeded one billion. Ethiopia leads the participating countries with an impressive figure of planting 700 millions trees followed by Turkey with 150 millions trees to Brazil at 16 millions. Indonesia launched a campaign to plant 79 million trees ahead of UN climate conference at Bali (Fig.15.3).

Afforestation Offset Projects

The Intergovernmental Panel on Climate Change (IPCC) reported in its third assessment that 10-30% of human-induced global GHG emissions are due to land use, land use change, and forestry. The IPCC concluded that globally, *changes in forest management could induce future carbon sequestration adequate to offset an additional 15-20% of* CO_2 *emissions*.(1b)

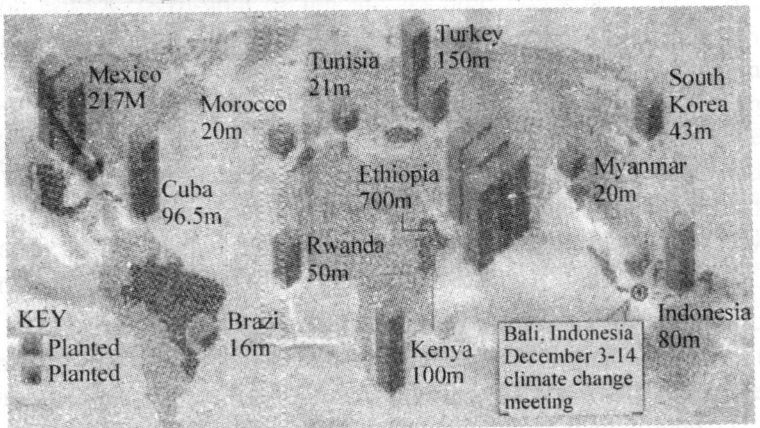

Fig. 15.3 Global Tree Planting in 2007 One Billion UN Environmental Program

Irreversible Environmental Changes by Deforestation

Deforestation not only cause the disappearance of trees but also loss of natural habitants and shrinkage of biodiversity. The majority of countries that have undergone massive deforestation are dealing with unforeseen transformations in climate and geography because the ability of forests to sustain a balance in climate changes is hindered.

"Avoided deforestation" is included in Bali Conference as a bold step in view of the mounting evidence that deforestation (despite carbon offset credit for afforestation in the current Kyoto regime) by destruction of tropical forests has caused as much as 20% increase in emission. At Bali, the countries, such as Brazil, wanted a system of reward for the nations, who have been able to cut their deforestation rates down to pre-1990 levels and keeping them there.

Manufacture of forest products

Afforestation by forest industries, by planting, managing and harvesting trees for making timbers and papers, to certain extent offset shrinking CO_2 sinks due to deforestation. Some of the entrepreneurs claim that their management system makes the project carbon neutral. Timber products generally have a longer life than paper; therefore can store carbon for longer, sometimes for centuries. However, most wood products are not so long-lived. Re-use and recycling prolongs the life of both timber and paper and can therefore delay the release of stored carbon back into the atmosphere.

Agroforestry and Biomass Energy

Agricultural products for making bio-fuel are financially most attractive proposition. For example, for every dollar invested in oil-palm plantations. Twenty dollar is recouped in a few years, even without the one-off gain from selling timber. Even if palm oil price slip downwards, forest-conservation plans need to be robust to compete[5].

Prevention of forest fire

Wildfires are very common in many places around the world, including much of the vegetated areas of Australia, Western Cape of South Africa, and the forested areas of the United States and Canada. The climates of these areas are sufficiently moist to allow the growth of trees, but have extended dry, hot periods. Fires are particularly prevalent in the summer and autumn and during droughts when fallen branches, leaves, grasses and scrub can dry out and become highly flammable. This is especially the case in areas where eucalyptus is prevalent, as the oil of these species is extremely flammable. News reports have suggested that global warming has been increasing the intensity and frequency of droughts in many areas, creating more intense and frequent wildfires.

In addition to forestry and beneficial effect on climate change, the other potential benefits of forests and forestry include biodiversity conservation, soil stability, amenity /recreation, cultural value and their role in hydrological cycles. A pragmatic and economical place to start restoring the balance of our global carbon cycle is in our forests. With good stewardship, forests will continue to provide not only carbon sequestration, but wood products and many other benefits, such as fish and wildlife habitat, biodiversity, clean water and recreation opportunities. Without good forest stewardship, we may lose the carbon battle and with it, most of the benefits our forests provide.

Indian scenerio on destruction of forest[6a]

The current status (2011) on forests of India can be summrized as follows:

i. Around 21% of surface area is covered by forests.

ii. Less than 31% of the forest cover is dense and rich in flora and fauna.

iii. The big loss of 1.47 lakh hectares during 1976 to 1980 has prompted government to enact the Forest Conservation Act in 1981.

iv. Destruction picked up in mid-2000 when the government set a target of 9% for economic growth. It reached the peak in 2009 and 2010, when 87,000 hectares of forest land were diverted (mainly for mining) - an area double to that diverted in previous five years.

The phenomenal growth of both legal and illegal mining industries is mainly responsible for destruction of forests, including the dense and pristine ones, with rich flora and fauna.

Carbon mapping in preserving biodiversity and forest[7]

An atlas correlating carbon and biodiversity has been produced by the World Conservation Monitoring Centre (WCMC) of the UN Environment Programme (UNEP). (See Chapter 6). Atlas maps those places that contain major species concentrations and where efforts to stop deforestation will produce maximum benefit. The atlas shows:

i High biodiversity areas within the tropical *Andes and Amazon account for 11 percent of the total carbon stock* in the area and have been named as *neotropics* by the experts.

ii. Africa *over 60 percent of the high biodiversity areas are in high carbon areas and contain a total of 18 billion tonnes of carbon.*

Employing the techniques used in the atlas would make it possible to identify where areas of high carbon density and high density of high biodiversity and thus ensure preservation of those biomes. For example,

i. In Tanzania, key biodiversity areas contain 17 percent of the country's carbon stock.

ii. Vietnam's protected areas cover 32 percent of the land area that has been identified as having high values for both carbon and biodiversity,

iii. In Papua New Guinea the map illustrates how the centre of the country, which is high in biodiversity, also contains large areas of high carbon stock. It also shows that existing protected areas overlap with only 14 percent of the high carbon areas.

The above findings demonstrate the potential value of the protected area system for meeting both carbon and biodiversity goals.

Technological carbon capturing & storing (carbon) will have their role, but the biggest & widest returns may come from investing in and enhancing 'natural carbon capture and storage systems.' In doing so, countries will forge part of a global green New Deal in which the infrastructure of economically-important ecosystems is renewed and renovated while sustaining livelihoods and hundreds of thousands of new green jobs in forestry and conservation in developing countries. Using the atlas, experts are looking at how investments in conserving carbon in the world could be a viable proposition.

vii. Afforestation Offset and Chicago Climate Exchange

Offsets are allowed for both additional removal of greenhouse gases and the avoidance of deforestation. In Chicago Climate market the offsets are allowed for the followings:

- Afforestation projects initiated on land that was degraded or bare as of January 1, 1990 and not required by law can earn CCX offsets.
- Afforestation projects that are implemented along with forest conservation can earn CCX.

A new market in forest carbon to provide revenue to landowners

The capacity of forests to become enhanced carbon sinks can bring added revenue to landowners through the emerging market in forest carbon credits. A well-organized forest carbon market can provide financial incentive for landowners to permanently conserve more forests and practice the type of management that results in carbon-rich forests.

In this market, forest owners committed to increasing carbon stores can sell these gains to entities seeking to offset carbon dioxide emissions. For example, the *Pacific Forest Trust* recently completed the first transaction in this emerging market through *Forest Climate Program* when it sold forest carbon credits to *Green Mountain Energy Company*.

By using existing scientific forest measurement tools, and ensuring that forest carbon projects yield permanent gains by securing them through conservation easements (also called a conservation covenant), forest carbon sequestration projects can be verifiable, enforceable, and provide carbon stores clearly additional to those that would have accrued otherwise.

Forest preservation-WWF efforts

For Ngoyle- Mintom forest in Cameroon-WWF proposes a comprehensive solution to maintain—through sustainable hunting and forestry to provide long-time livelihoods for locals, while preservation of core area pristine forest should protect species. In this process the needs of nature conservation and human are made compatible. This forested landscape would help avert damaging climate change[8].

Ecotourism[9]

Ecotourism is one of the few non-destructive land uses, preserving the biome, with capability of generating an immediate, competitive cash flow, along with economic and social development of the region. Peru's Madre de Dios region has been undergoing an ecotourism boom[8]. A large number of 'eco-lodges' – thatched 30-bed tourist lodges, built in a region with vast areas of pristine rainforest, including some of the most biodiverse places on the earth, much of it protected in magnificent natural park. Rainforest expeditions, run by the biggest tourist operator in the region, have entered into a twenty year joint venture with the local community, who shares the decision making through an elected 'committee', and receive 60% of the profits, amounting to $130,000 last year. They also got most of the last year's $140,000 payroll. The company has been able to undertake conservation, social and economic development more nimbly than government or NGOs.

Ocean and Global Warming

Marine regions cover about three-fourths of the Earth's surface and include oceans, coral reefs, and estuaries. Ocean covers 97 percent of marine regions while rest 3% belongs to freshwater. The world's oceans have an even greater effect on global climate than forests do.

The total amount of carbon in the ocean is about 50 times greater than the amount in the atmosphere, and is exchanged with the atmosphere on a time-scale of several hundred years. At least 1/2 of the oxygen we breathe comes from the photosynthesis of marine plants. Currently, 48% of the carbon emitted to the atmosphere by fossil fuel burning is sequestered into the ocean. But the future fate of this important carbon sink is quite uncertain because of potential climate change impacts on ocean circulation, biogeochemical cycling, and ecosystem dynamics[10]. In addition to fresh water, ocean stores heat, salt and carbon dioxide, and circulates these necessities around the planet's surface. The water, carbon dioxide and the heat in the ocean – all play important role in global warming.

Heat storing and transport by ocean

Ocean's fundamental role in climate is based largely on their storage and transport of heat around the globe, thus keeping atmospheric temperature fairly constant. The oceans store vast amounts of heat, much more than the heat stored by the atmosphere, because water is 1000 times denser and has a heat holding capacity (heat capacity) four times that of air. The *ocean can store about thousand times more heat than the atmosphere.* Ocean currents are primary highways for the transport of heat around the globe. The atmosphere, with less heat content moves faster and thus can heat around the world much more quickly. Without the warm ocean currents transferring heat to the atmosphere, certain parts of the globe, such as northern European countries including Britain, would not have the temperate climate.

The carbon and hydrologic cycles are the two natural processes regulating the global warming and climate changes. The carbon cycle involving natural exchange of carbon by ocean water with the atmosphere, determines the quantum of extra carbon in the atmosphere and thus the extent of global warming. The hydrologic cycle determines the humidity of the atmosphere and precipitation (rainfall) and thus has a significant effect on global temperature and climate. Both the natural processes are affected by anthropogenic emissions making the cycle unbalanced. To cope up with the excess emissions it is essential to retain and improve the carbon absorption capacities of oceans. To preserve the carbon exchange capacity it is necessary to prevent pollution detrimental to marine life and to stop eliminating already 'endangered species'. In addition to these measures, cultivation of phytoplankton can cause an increase in the capacity to absorb carbon (Chapter 6). The carbon cycle and the hydro cycle are interrelated (Chapter7) and thus both play important role in global warming.

Ocean-atmosphere hydrologic cycle

The ocean contains 97% of the fresh water on Earth, experiences 86% of evaporation, and directly receives 78% of precipitation. The evaporation of water from ocean surface and return by precipitation complete the hydrologic cycle. Together, the atmosphere and ocean maintain a global balance in the distribution of fresh water (Chapter 7).

Evaporation rates should increase as the surface ocean heats up. Also, because water vapor pressure rises exponentially as temperature increases, a warmer atmosphere will hold more water vapor. Since water vapor is itself a potent greenhouse gas (much more so than carbon dioxide), increased water vapor concentrations will trap additional heat and cause the surface temperatures to rise even faster. Therefore the higher evaporation rate in hydrologic cycle would result because of global warming, and that this will in turn shall accelerate global warming.

Ocean-atmosphere carbon cycle

In the long term, the ocean plays the dominant role in the natural regulation of CO_2 in the atmosphere and thus exerts a powerful influence on the climate. The *natural carbon exchange of ocean* with atmosphere takes place by following three processes *(see chapter 7)*

i. By *physical pump* which is driven by gas exchange at the air-sea interface and the physical processes that transport CO_2 to the deep ocean.

ii. By *biological pump i.e.* phytoplankton taking up nutrients and CO_2 through the process of photosynthesis.

iii. The ocean "breathes" *i.e.*, ocean uptake and out gassing of CO_2 across the sea surface. Hot water gives up CO_2 while cool water absorbs CO_2.

Ocean plays a vital role in the Earth's carbon cycle and currently sequesters up to 48% of the carbon by burning fossil fuel. A recent report[11] indicates the quantum of carbon pool and fluxes in ocean as follows:

Carbon Pools in ocean

1. Surface Water	1,020 billion tons
2. Intermediate & deep water	38,100 billion tons
3. Oil & gas	300 bt
Carbon flux between ocean and	
Out of solution	90 billion ton
Into solution	92 billion ton

[$1000^3 = 10^9$ = giga = Billion]

Net Carbon flux remains same as 2.0 Gt in 1900s as given in table 15.1

Carbon dioxide uptake by ocean can be improved by keeping the normal activities of the nature's physical and biological pumps.

Physical pump

Atmospheric CO_2 enters the ocean by gas exchange depending on wind speed and the differences in partial pressure across the air-sea interface. The amount of CO_2 absorbed by seawater is also a function of temperature through its effect on solubility. *Solubility increases as temperatures fall so that cold surface waters pick up more CO_2 than warm waters*. The solubilty of carbon dioxide in the sea water can be increased by limiting the rise of global temperature.

Biological Pump

The rate at which photosynthesis process occurs is called the primary productivity. Some of the organic matter created is cycled through the food web in the upper ocean and some sinks to the bottom. Some of this carbon is rematerialized back to CO_2 while a tiny fraction is buried in the sediments of the sea floor. *Biological pump* is run by planktons through photosynthesis. They are capable of making the process self sustaining with the help of sunlight. Phytoplanktons convert water, minerals, and carbon dioxide under sunlight into carbohydrates and oxygen

For photosynthesis it is necessary to have sun light and minerals. Physical processes such as upwelling, hydrographic fronts, eddies and cyclones bring nutrients from deeper waters to the ocean surface. The top 100-150 m of the ocean surface (photic or euphotic zone) as well as shallow water zones in the coastal areas get the required level of light intensity for photosynthesis. These zones are thus favorable for producing phytoplankton. Tropical and sub-tropical oceans with abundant light have low primary production because of poor availability of nutrients. In such regions, primary production occurs at greater depth, although at a reduced level (because of reduced light). (ref. Chapter6)

Recently there has been a growing commercial interest in the seeding process, because of the possibilities of selling carbon credits worth millions of dollars. Much of the Southern Ocean is depleted of iron and experiments have shown even a small amounts of nutrient can trigger phytoplankton blooms that can last for up to two months. Some entrepreneurs have started iron seeding on commercial scale in icy seas between Australia and Antarctica for producing plankton which could become a money spinner.

Oceans are naturally alkaline, or basic, with a pH of about 8.2. When carbon dioxide dissolves in sea water it forms carbonic acid (H_2CO_3), making it more acidic. The high acidity can be harmful for planktons to grow. Iron seeding can increase the growth pf phytoplankton and thus speed up absorption of carbon dioxide by phytoplankton instead of sea water. By 'fertilizing' the ocean with iron would also prevent ocean acidification.[12]

Human activities destroying ocean biome

Oceans for centuries have been considered as resource to provide inexhaustible supply of food, useful transport route, and a convenient dumping ground. But human activity, particularly over the last few decades, has finally pushed oceans to their limit. *Pollution, habitat destruction, unsustainable fishing, tourism development, oil and gas extraction, shipping, and aquaculture are all taking their toll on marine habitats, marine species and people.* These factors would also affect hydrologic cycle and eventually the climate. Ocean water circulation (chapter2, p25) systems, which controls excessive temperature rise depends on the density and salinity of the water get badly affected by pollution. *Pollution affects density and salinity thus carbon dioxide sinking ability.* This would have not only an adverse effect on conveyor belt type circulation but also spread of pollution.

Prevent Marine pollution

Over 80% of marine pollution comes from land-based activities. Most of the waste produced on the land eventually reaches the oceans, either through deliberate dumping or from run-off through drains and rivers. These include, oil, fertilizers, solid garbage, sewage, and toxic chemicals

Oil spills include releases of crude oil from tankers, offshore platforms, drilling rigs and wells, as well as spills of refined petroleum products (such as gasoline, diesel) and their by-products, and heavier fuels used by large ships such as bunker fuel, or the spill of any oily white substance refuse or waste oil. Oil spills cause huge damage to the marine environment. An estimated amount of around 12% of the oil enters the seas each year. According to a study by the US National Research Council, 36% comes down drains and rivers as waste and runoff from cities and industry[13]. *Fertilizer runoff* from farms and lawns is a huge problem for coastal areas. The extra nutrients cause eutrophication - flourishing of algal blooms that deplete the water's dissolved oxygen and suffocate other marine life. Eutrophication has created enormous dead zones in several parts of the world, including the Gulf of Mexico and the Baltic Sea. *Sewage disposal* flows untreated, or under-treated, in many parts of the world into the ocean. For example, 80% of urban sewage discharged into the Mediterranean Sea is untreated.

Plastic garbage, which decomposes very slowly, is often mistaken for food by marine animals. High concentrations of plastic material, particularly plastic bags, have been found blocking the breathing passages and stomachs of many marine species, including whales, dolphins, seals, puffins, and turtles. Plastic six-pack rings for drink bottles can also choke marine animals. *Toxic chemicals* dumping at sea include material such as pesticides, chemical weapons, and radioactive waste. *Toxic waste* gets into seas and oceans by the leaking of landfills, dumps, mines, and farms. Farm chemicals and heavy metals from factories can have a very harmful effect on marine life and humans. Harmful chemicals in the ocean are lead & mercury. *Mercury* in the oceans has increased 5 times since the industrial revolution and is getting worse, with an estimated 2% rise in mercury per year. Top ocean predators like sea mammals and tuna have high levels of mercury in their meat, ranging from 5 to 3500 times the amount allowed by Japanese law. Mercury originates mainly from the burning of fossil fuels (coal and gasoline). The releases mercury from combustion of coal accounts for 70 percent of this pollutant's presence in the atmosphere. To reduce the mercury in the oceans it is essential to reduce use of coal as a source of energy.

The man-made pollution of ocean is controllable. Several countries have taken steps to reduce pollution of the sea.

Summary

i. Both forest and ocean have large cabin storage capacity (carbon pool) and absorbs substantial quantities of carbon as carbon flux in their interaction with atmosphere. They act as major carbon sink along with the biodiversities.

ii. Human activities in recent decades have caused enormous harm to these biomes, which has resulted in the reduction in the capacity to absorb carbon from the atmosphere. The net result is the accumulation of anthropogenic carbon in the atmosphere with potential to increase global average temperature to an alarming scale.

iii. Prevention of deforestation and reforestation are major steps to restore the carbon storage capacity of forest.

iv. Pollution control, preserving marine life and salinity of water, preserving and increasing population of phytoplankton by seeding are the major steps required to maintain and increase the carbon absorption capacity of oceans.

REFERENCES

1.a. *The supplementary report to the IPCC scientific assessment1992.*

1.b. IPCC Fourth Assessment, 2007, Cambridge University Press.

2. Sandra Brown, *Present and potential roles of forests in the global climate change debate*, US Environmental Protection Agency, Western Ecology Division, Oregon, USA,http://www.fao.org/ docrep/ w0312e/w 0312e03 .htm).

3. ARM Program Upgrades Eddy Correlation Flux Measurement Systems to measure Carbon, ARM Climate Research Faculty, DOE, USA http://www.archive.arm.gov/Carbon/upgrades.html

4. Earth Trends (2003). "Biodiversity and Protected Areas - Costa Rica" (PDF). World Resources Institute;.http://earthtrends.wri.org/ df_library/country_profiles/bio_cou_188. pf. Retrieved 2008-06-08.

5. Marland and Marland, Should we store C in trees? *Water Air Soil Pollut.* 64: 181-195. 1992.

6. *The Economist*, December, 22nd, 2007, p98

6a. Arvind Samant, Knock on Wood, *Hindustan Times*, Mumbai edition, October 2011.

7. *Biodiversity Demarcation Atlas of Biomes*: New atlas from UN Environment Programme /Adobe PDFwww.unep.org/pdf/ carbon_biodiversity. Pdf

8. James Leape, Director General, WWF International, Gland, Switzerland, p18, letters to Editor, The Economist, 8th march, 2008, p18, Main article in Economist, feb,16th, 2008.

9. *Economist*, April 12, 2008, p 50.

10. NASA report on Ocean Carbon Cycle http://science.nasa.gov/earth-science/oceanography/ocean-earth-system/ocean-carbon-cycle/

11. Carbon Pool & Fluxes in Ocean, ACECRC, Wetland International, 2008.

12. Nova Science in the news, published by Australian Academy of Science in website; http://www.science.org.au/nova/106/106print.htm

13. Oil Spill, Wikipedia.

17

Agriculture and Emission Management

While the increasing concentrations of GHGs are associated primarily with fossil fuel consumption, a significant share (estimated in the range of 12 to 42 per cent) is believed to be caused by changes in land use, including deforestation and the expansion of agriculture. The GHGs can be reduced in the atmosphere by reversing some of the processes associated with land use changes and adopting appropriate agriculture management practices.

Agriculture and GHG Emission

Agriculture is a significant emitter of GHGs, particularly methane, nitrous oxide, and CO_2. Thus policies designed to reduce emissions need to be targeted to these agricultural sources. The IPCC (1996) estimates[1,2] that globally agriculture emits about:

* 50 per cent of total methane, sources include rice, ruminants and manures;
* 70 per cent of nitrous oxide, sources include manure, legumes and fertilizer;
* 20 per cent of CO_2, sources include use of fossil fuels, soil tillage, deforestation, biomass burning, and land degradation.

Emissions can vary substantially, between developing and developed countries. Deforestation and land degradation mainly occur in developing countries. Agriculture in developed countries uses more energy, more intensive tillage systems and more fertilizer, resulting in fossil-fuel based emissions, reductions in soil C and emissions of nitrous oxides. In addition animal herds emit methane from ruminants and manure.

Agriculture and Land use Change as Source of Carbon Flux

Agriculture and forests are mentioned as both emitters of and sink for GHGs in the Kyoto Protocol. Annex A of the Kyoto Protocol lists agriculture as an *emission source* in terms of:

i. enteric fermentation,

ii. manure management,

iii. rice cultivation,

iv. soil management,

v. field burning, and

vi. deforestation.

In agriculture, carbon flux is mainly due to methane (CH4) & nitrous oxide (N_2O). Carbon flux of CO_2, CH4, & N_2O, can occur due to land-Use Change and Forestry as follows:

- Forest Carbon Flux
- Liming of Agricultural Soils
- Urban Trees
- N_2O from Settlement Soils
- Non- CO_2 Emissions from Forest Fires
- Landfilled Yard Trimmings and Food Scraps

The Kyoto Protocol also lists agriculturally related *sinks* of afforestation and reforestation. Additional sources and sinks which are under consideration include agricultural soil carbon and water. Some of the important sources of GHGs in agriculture and the ways to minimize emissions are to be discussed in this chapter. Emissions of GHG from waste food and related control are also included.

i. Enteric Fermentation

Fermentation that occurs in the digestive systems of ruminant animals is termed as enteric fermentation. It is one of the factors responsible for increased methane emissions. Ruminant animals are those that have a *rumen,* a special stomach found in cows, sheep, and water buffalos that enables them to eat tough plants and grains which monogastric animals, such as human, dog and cat cannot digest. Enteric fermentation occurs when methane (CH_4) is produced in the rumen as microbial fermentation takes place. Over 200 species of microorganisms are present in the rumen, although only about 10% of these play an important role in digestion. Most of the methane so produced in the rumen is belched by the animal. In Australia *ruminant animals account for over half of their green house gas contribution from methane*. Australia has implemented a voluntary immunization program for cattle in order to help reduce methane production in rumen.

ii. Manure Management-manure as resource

Using Manure as a Nutrient Source in Crop and Garden Production: The separation of animal agriculture from crop production has led to accumulation of excess manure on livestock farms. Crop farms can benefit from this manure as a source of nutrients and organic matter, if the manure is suitable for their needs and shipping does not make the cost prohibitive. Nutrient values of different sources of manure are being assessed with respect to their suitability for crop production. Guidelines to use manure as fertilizer for crop farmers are available.

Horse manure and soil nitrogen: Horse manure is an abundant, locally available source of organic matter for soils. A major concern about horse manure is that it can cause a nitrogen deficiency when added to soils, leading to stunted, yellowed crops.

Managing Dairy Manure

Water quality problems, changing herd management patterns, and increased regulation have made manure management a critical issue for dairy farmers. The immediate goal is to help dairy farmers improve the use of manure to increase agronomic benefits and reduce the risk of over-application, runoff, and leaching. Manure application rates have traditionally been based on nitrogen, but phosphorus has emerged as the nutrient of concern in many watersheds.

iii. Rice Cultivation

Methane Emissions from Flooded Rice Cultivation

By 2020, the world will need to produce 350 million tons more rice per year to feed an anticipated 3 billion more people than in 1992. The rice field methane emissions have been identified as a major source of atmospheric methane. During flooding of the field in wetland rice soils, the oxygen supply from the atmosphere stops and leads to anaerobic fermentation of soil organic matter. Methane, a major end product of anaerobic fermentation, is released from submerged soils to the atmosphere through the roots and stems of rice plants. Estimates of global methane emission rates from rice fields range from 20 to 100 Tg per year (1 Tg=1 million tons), which corresponds to 6 to 29 percent of total annual anthropogenic methane emission.

Methane emission from rice field was found to be affected by various factors some of which are as follows:

- Exogenous organic matter appears to be the largest contributor to methane production from flooded rice soils.
- In coastal rice fields of Texas, it was found that as the rice growing season progressed, methane production at lower depths and farther away from the plants increased in proportion with root density.

iv. Agricultural Soils: Source of Emission & Carbon Sequestration

In some places (*e.g.* North Carolina), soils are naturally acidic and need lime, which neutralizes the acidity, for optimum growth of crops, forages, turf and trees. Soils become more acid due to the leaching of calcium (Ca2+) and magnesium (Mg2+), or acidification by hydrogen added to soils by decomposition of plant residues and organic matter and during the nitrification of ammonium added to soils as fertilizer. Agricultural lime (Agilime) includes crushed limestone ($CaCO_3$) and dolomite (Ca-,Mg- CO_3). Following IPCC that all carbon in agrilime is eventually released as CO_2 to the atmosphere, the US-EPA (2a) estimated that 9 Tg ($1 Tg = 10^{12}$ g = 10^{6} metric tons) of CO_2 was emitted from an application of 20 Tg of aglime in 2001.

Globally, soils are estimated to contain approximately 1,500 gigatons of organic carbon, more than the amount in vegetation and the atmosphere. Modification of agricultural practices is a recognized method of carbon sequestration as soil can act as an effective carbon sink offsetting as much as 20% of 2010 carbon dioxide emissions annually[3].

Scientists estimate that about 80 percent of global carbon is stored in soils, and that a substantial proportion of carbon that was originally in soils has been released due to human land use, implying that there is a large technical potential to sequester carbon in soils Agriculture may enhance the soil's capacity to absorb GHGs by creating or expanding sinks. This may be achieved through a variety of changes in land use and management practices. The adoption of agroforestry practices like windbreaks and riparian forest buffers, which incorporate trees and shrubs into ongoing farm operations, represents a potentially large GHG sink nationally. In addition, agricultural practices such as conservation tillage and grassland practices such as rotational grazing can also reduce carbon losses and promote carbon sequestration in agricultural soils. These practices offset CO_2 emissions caused by land use activities such as conventional tillage and cultivation of organic soils[4]. The carbon sequestration management processes should basically cover the followings:

- land retirement (conversion to native vegetation or reversion to wetlands),
- afforestation,
- residue management,
- less-intensive tillage,
- changes in crop rotations,

- conversion of cropland to pasture and restoration of degraded (or highly eroded) soils.

Some of these management practices are discussed in details in the following paragraphs.

Agricultural Land Use Changes

The agricultural land use changes to cater for the increasing demand of bio-fuels can result in increased carbon emission thereby negating the climate change mitigation programme. Recent Oxfam studies[4] indicate that by 2020, carbon emissions resulting from land use changes in Europe because of rising demand of bio-diesel derived from palm oil would be between 3.1 to 4.6 billion tonnes, which is 46 to 68 times higher than the EU hopes to be achieving by then from bio-fuels. Also the carbon emissions from global land use changes due to US corn-ethanol programmed will take 167 years of climate mitigation programmed to pay back. In addition to increased emission the land use changes for biofuel has also resulted in an estimated increase in food price by 30 per cent[5].

However the proper management practice can make the biofuel as carbon neutral. Burning biomass based fuels not only reduces carbon emission in comparison to fossil fuels, but also can reduce net CO_2 emissions to nearly zero, if the growing plants can absorb equivalent carbon in their photosynthesis process for biomass growth (Chapter 12).

No-till soil management[6]

During tilling. soil is turned, the organic matter is exposed to the atmosphere and can quickly oxidize into carbon dioxide. This is bad for the soil and is thought to be an important emission source driving global warming. Less organic matter in the soil means less water retention, less nutrient release and less clod formation. Along with encouraging the oxidation of organic matter, plowing physically breaks up pre-existing clods (lumps) and exposes them to the direct force of rainfall. The force exerted by rain drops is enough to break up the remaining clods and form a structureless soil "crust" on the field surface. This crust has virtually no pores and as a result, plant roots do not get the water and oxygen that they need, and surface runoff and soil erosion become a large problem. The crust can also be strong enough to make it difficult for seedlings to push through to the surface. In no-till agriculture, the farmer uses a disk or chisel plow to prepare the field for seeding. Rather than turning the field, these plows create a narrow furrow, just large enough for the crop's seeds to be injected. Tractor attachments inject a band of fertilizer in with the seeds, thus negating the need to fertilize the whole field, and close up the furrow after the seed and fertilizer have been planted. With these new plows, the farm field can be seeded with minimal disturbance of the soil.

Advanced field management coupled with modern no-till plowing can actually increase the fertility of a farm's soil. The use of cover crops and green manures during the non-growing season are two popular forms of such field management. A cover crop is any sort of vegetation that is grown between commercial plantings. The cover crop holds the soil in place with its roots, preventing erosion. Green manures are any type of crop that is incorporated into the soil while still green or soon after flowering. In many cases, cover crops become green manures just before commercial planting resumes.

The benefits of no-till farming are economic as well as environmental. The no-till farmer will see an increase in the organic matter of the soil, and a decrease in the amount of erosion. More organic matter and less erosion mean more fertility, less fertilizer, and higher yields. Additionally, with the advances in cover crops and green manures, the no-till farmer can greatly reduce the use of high-cost herbicides.

According to Prof Johan Rockstrom, an environmental scientist and the director, Stockholm Environment and Stockholm Resilience Centre[6], India with its agro-base economy has around 30% of carbon emission from old farming technique of ploughing and need to go for no till practice. Some Latin American countries have gone for no-till practice. In USA, around 40% of farming have been using no-till practice.

Carbon savings through management practices

The various management practices can lead to following percentages in carbon savings:

- 49 per cent of agricultural carbon sequestration can be achieved by adopting conservation tillage and residue management,
- 25 per cent by changing cropping practices,
- 13 per cent by land restoration efforts,
- 7 per cent through land use change and
- 6 per cent by better water management.

Summary

Agriculture is a significant emitter of GHGs, particularly methane, nitrous oxide, and CO_2. Methods used to limit emission in agriculture inlude followings:

- Reduction of methane emission by immunization against enteric fermentation.
- Increase soil carbon by less-intensive tillage, residue management, changes in crop rotations, conversion of cropland to pasture and restoration of degraded (or highly eroded) soils.
- Land use change to make biofuel can have disastrous consequences on environment and global warming, if not managed to make carbon neutral.

REFERENCES

1. IPCC (Intergovernmental Panel on Climate Change). 1996a. Houghton, J., L. Meira Filho, B. Callander, N. Harris, A. Kattenberg, and K. Maskell, (eds.). *Climate Change 1995: The State of the Science.* Cambridge University Press.
2. IPCC (Intergovernmental Panel on Climate Change). 1996b. Watson, R., M. Zinyowera, R. Moss, and D. Dokken (eds.) *Climate Change 1995: Impacts, Adaptations, and Mitigation of Climate Change: Scientific-Technical Analyses* Cambridge University Press

2a. EPA, 2004; Inventory of US Greenhouse Gas Emissions and Sinks, 1990-2002,EPA430-R-04-003. US Environmental Protection Agency.

3. U.S. Agriculture and Forestry Greenhouse Gas Inventory: 1990-2001, http://www.usda.gov/oce/climate_change/inventory_1990_2001/USDA%20GHG %20Inventory% 20Chapter%201.pdf

4. Carbon Sequestration; Wikipedia, http//en.wikipedia.org/wiki/Carbon_seqestration

5. Another Inconvenient Truth. *How biofuel policies are deepening poverty and accelerating climate*, Oxfam Studies, June, 2008.

6. Report in Hindustan Times, Mumbai edition, Nov03, 2009, p04, India.

18

Waste Management

Waste management refers to the process of (i) collection of waste matter generated mainly by human consumption and activity, (ii) transport and shipment of the collected waste matter to a waste treatment facility and (iii) processing/recycling this waste material for further use or disposing it for good. Waste can be in the form of solid, liquid or gas.

Before industrialization, waste generated was minimal and manageable. Human wastes were mostly biodegradable with minimum impact on the environment. With the increase in population and the industrial revolution, human consumption began to get concentrated and waste began multiplying. The growing quantities of industrial wastes have led to high level of pollution of water, soil and air, affecting natural cycles and biomes. The crippling effects of pollution on important sinks like forest, soil and ocean have reduced their capacities to absorb GHGs. The waste management processes should therefore include the recycling of recovered valuable from waste, processing the waste for reuse and making the waste eco-friendly before dumping. The waste management thus plays the dual role of decreasing GHG emissions and preserving the capabilities of sinks to absorb GHGs. EPA estimates that simply increasing our national recycling rate from its current level of 30 percent to 35 percent would reduce GHG emissions by another 10 million tons of carbon equivalent (MTCE)[1].

Life Cycle of Product and GHG

The different stages of a product's life cycle generate waste till the end when the product itself becomes a waste. A product's life cycle can include following stages:

i. **Extraction:** The primary activity of extraction of minerals in mining leads to generation of unwanted materials associated with the required minerals as waste.

ii. **Manufacturing:** The manufacturing process is normally accompanied by certain amount of waste along with the end product. The unavoidable processing waste generated can be more than one type. For example, in iron and steel making plant, the list of waste products includes solid (slags), liquid (pickling liquor), and gas (Blast Furnace gas). However the iron & steel industry has been able to effectively manage to convert some of the waste into resources or make it eco-friendly before disposal. Thus the slags are used for slag cement, the heat energy of BF gas is recovered through recuperators, and the pickling liquor is neutralized before discharge. The avoidable process waste can be minimized by effective control of inputs and the process parameters.

iii. **Transportation:** Transportation of intermediate, final products and waste by vehicular transport leads to GHG emissions. The utilisation of waste and waste treatment facilities in side plant can reduce the tranportation activities significantly. In steel melting plants using arc furnace or open hearth process, the steel scraps produced inside the plant are recycled for production of fresh materials. Pickling liquor treatment facility inside the plant allows it to discharge locally.

iv. **Packaging and Distribution:** The distribution of products need packaging and thus eventually generates packaging waste. According to Waste Resources Action Programmed (WRAP, UK), an estimated 6.8 Mt of household food waste, equivalent to one third of all purchased food, is produced annually in the UK[2]. This represents over 8 billion pounds in retail value and costs families on average 250-400 pound a year. Assuming 50% of the food could have been eaten, avoidable waste equates to at least 15 Mt. Packaging protects against damage, contamination and spoilage and printing on the pack provides information to customer. Data from US municipal solid waste studies show an inverse linear relationship between waste food and packaging residues.

v. **Worn equipments and critical components as scrap (waste):** Worn engineering components, like gears, shafts, bearings etc, become scrap or waste at the end of the life cycle or due to premature failure. Some of the critical components failure may make the whole equipment or machinery as scrap or waste. Increase in life cycle of engineering components forming vital parts of equipments, machineries, cars, jet planes etc., prevents the whole or major critical components of the equipment from getting scraped as waste. The life cycle improvement not only prevents generation of waste but also leads to saving of materials and energy required to make new products as replacement to scrapped ones[3].

As manufacturer or user of a product, it is useful to analyze the entire product life cycle to determine where organization can make changes, such as preventing waste, recycling, or buying or manufacturing recycled products, in order to reduce the carbon footprint and thus its impact on global warming. For manufacturers, life cycle analysis provides opportunities for producing goods using less material, which means that less energy is needed for extracting, transporting, processing raw materials and transporting end products. Manufacturing goods from recycled materials is beneficial because it requires less energy than producing goods from virgin materials. Also refurbishing of worn equipments and critical components save enormous amount of wastage and leads to reduction in carbon footprints of the components and the equipment[4].

vi. **Disposal:** Disposal of untreated waste can generate GHGs or cause pollution leading to destruction of biomes. Disposal of organic materials like food, paper, and yard waste in landfills can lead to methane formation and subsequent emission from the landfills. Untreated waste can cause havoc by destroying the large number of species living within the affected region.

Waste Management Techniques

The waste management techniques include the followings:

1. **Source reduction and Conservation:** Waste prevention, or "source reduction," means consuming and throwing away less. The necessary steps to waste prevention should include, purchasing durable, long-lasting good, seeking products and packaging that are as free of toxics as possible and redesigning products to use fewer raw materials in production, have a longer life, or be used again after its original use. Source reduction actually prevents the generation of waste in the first place, so it is the most preferred method of waste management and goes a long way toward protecting the environment.

2. **Recovering Resources from Waste:** As the world population increases and waste grows in volume, the world's scientists and planners have evolved technologies to recover resources from waste, which can be used again. For example, the developed nations have sophisticated facilities that convert the calorific content present in waste into electricity. In developing nations, manual laborers sift through the waste and extract recyclable material from it, thereby reducing the volume of waste that needs to be disposed.

3. **Recycling:** Recycling involves processing of used or worn materials into new products to prevent waste of potentially useful materials. Recycling reduces the consumption of fresh raw materials, reduce energy usage, reduce air pollution (from incineration) and water pollution (from land

filling). Recycling of scraps in manufacturing reduces the need for "conventional" waste disposal and requires less energy in comparison to that of the virgin production. Recyclable materials include many kinds of glass, paper, metal, plastic, textile and electronic items.

Recycling is a key component of modern waste management and is the third component of the 3-R principle, viz., "Reduce, Reuse & Recycle" in waste hierarchy. In the case of worn machine components, recycling are carried out after repair and reconditioning and the refurbished equipment life span can be more life than original[5]. Electronic waste is sent to developing nations where recycling plants extract gold and copper from the e-waste. Used automobiles are scrapped and their metallic parts are sold to factories for re-conversion and so on. Stainless steel, made from at least 60% recycled scrap is one of the most recycled materials. Even after lasting for 100 years or more the stainless steel parts can be recycled again to make 'new stainless steels[6]. Collecting and processing secondary materials, manufacturing recycled-content products, and then purchasing recycled products creates a circle or loop that ensures the overall success and value of recycling. If waste cannot be recycled, *incineration* or *sanitary land filling* is the next preferred methods of treatment.

4. **Landfill:** This is the most traditional way of managing waste, by dumping it in a lan 'Tll. There are several types of solid waste landfills, such as, municipal solid waste, construction, demolition and industrial waste.

   Modern landfills are well-engineered facilities that are located, designed, operated, and monitored to ensure compliance with rules and regulations, formulated by competent authorities. Solid waste landfills must be designed to protect the environment from contaminants which may be present in the solid waste stream. The *landfill sitting plan*—which prevents the sitting of landfills in environmentally-sensitive areas—as well as on-site environmental monitoring systems—which monitor for any sign of groundwater contamination and for landfill gas—provide additional safeguards. In addition, many new landfills collect potentially harmful landfill gas emissions and convert the gas into energy.

5. **Incineration:** This involves the disposal of waste by burning it. However, incineration is not an effective tool for waste management, excepting for hospital waste. The burning of waste consumes resources and energy, destroys the recyclable material present in the waste and emits many harmful pollutants.
6. **Composting:** This is a technique in which organic waste materials (food, plants, paper) are decomposed and then recycled as compost for use in agriculture and landscaping applications. Compost is defined as relatively stable humus material that is produced from a composting process. In the composting process bacteria in soil convert biodegradable waste in the garbage into organic fertilizer. The natural process of composting is to pile up the waste outdoors for a year or more in order to allow the degradable trash to break down into organic fertilizer. Modern composting is a multi-step, closely monitored process with measured inputs of water, air and carbon- and nitrogen-rich materials. The decomposition process is aided by shredding the plant matter, adding water and ensuring proper aeration by regularly turning the mixture. Worms and fungi further break up the material. Aerobic bacteria convert the inputs into heat, carbon dioxide and ammonium. The ammonium is further converted by bacteria into plant-nourishing nitrites and nitrates fertilizers through the process of nitrification.

The compost itself is beneficial for the land in many ways, such as. soil conditioner, fertilizer, and natural pesticide for soil. Compost is a biomass and therefore can be used to produce biofuel.

This is one of the most effective methods of reducing the amount of material in the waste stream. Yard trimmings and food residuals together constitute 24 percent of the U.S. municipal solid waste stream. That's a lot of waste to send to landfills when it could become useful and environmentally beneficial compost instead.

7. **Mechanical Biological treatment:** In this technique, a variety of waste (plastic, paper, glass, etc.) are fed in bulk into the waste treatment plant. The MBT process extracts the recyclable content in the waste and converts it to calorific fuel. The heat source produced by MBT can be gainfully used by cement/power plants.
8. **Pyrolysis and Gasification:** These are thermal techniques, where waste is treated at high temperatures and at a very high pressure. In pyrolysis, the waste material is converted to solid or liquid. The solid material can be further refined into a carbon form while the liquid extract can be used as energy-giving oil. In gasification, the waste material is converted into a synthetic gas, which can be burned to produce more energy[7].

Liquid Waste Management

Liquid waste disposal without pretreatment can be a potential danger to GHG sinks, such as, soil, water, and forests. The important aspects of liquid waste management include the followings:

i. **Source control and pre-treatment:** The source control and pretreatment are essential steps to reduce the organic load, toxicity and volume of industrial commercial waste. The quality of discharge to sewers should conform to stipulations made by regulatory agencies in their specifications.

ii. **Recycling:** Recycling and utilizing waste materials can have long-term economic and social benefits. For instance, it may be preferable to treat the sewage in satellite plants for reuse as irrigation water on surrounding forests, farm land, or community facilities such as parks, golf courses and boulevards, even though this may incur additional costs. Similar arguments can be made for recycling sewage effluent for its nutrient content, or recycling sludge for its humus and nutrient content.

Waste as Energy source

Gasification process is used to convert biomass or organic waste into carbon monoxide and hydrogen by controlled combustion with oxygen. to form fuel called *synthetic gas or syngas*. (see chapter on Chapter 12)

Anaerobic digestion is a process to make biofuel, in which microorganisms break down biodegradable materials in the absence of oxygen. The process is widely used to treat wastewater sludge and organic wastes, such as waste paper, grass clippings, leftover food, sewage and animal waste. except woody waste because of non-degradable lignin. Anaerobic digestion is a *renewable energy* process producing methane-rich biogas suitable for energy production. Also, the nutrient-rich solids left after digestion can be used as *fertilizer.* As part of an integrated waste management system, anaerobic digestion reduces the emission of landfill gas into atmosphere. However, technical expertise required to maintain anaerobic digester plus high capital cost and lower efficiencies have so far limited this vital waste treatment technology, although recognized by UNDP as one of the most useful decentralized sources of energy supply, as they are less capital intensive than large power plants. From 1975, China and India have large government-backed schemes for adaptation of small biogas plants for use in the household for cooking and lighting. Presently, projects for anaerobic digestion in the developing world can gain financial support through the United Nations Clean Development Mechanism for reduced carbon emissions (chapter 9 & 10).

Biorefinery

Biorefinery can produce multiple products including petrol and diesel. Most promising is cellulosic ethanol. Cellulosic ethanol can be made from inedible cellulose fibers that form the stems and branches of plants. Crop residues (such as corn stalks, wheat straw and rice straw), wood waste, and *municipal solid waste* are potential sources of cellulosic biomass (see chapter12)

Waste Water to Grow Seaweed for biofuel

In recent years seaweeds have been used to produce ethanol, for use as biofuel. Apart from natural seaweeds aquaculture technique is used to grow seaweeds. However, millions of tonnes of untreated wastewater are dumped daily into seas and seaweed helps clean it up. Therefore growing large seaweed fields for energy by using waste water as nutrients is a sound economical proposition. This idea has been tested successfully using human waste water in experiments at US institutions, including the Woods Hole Oceanographic Institution and Harbour Oceanographic Institution (see chapter12).

Summary

i. The management of waste includes the processes under '3R' (reduction, reuse, recycling) as described earlier, plus those under '2R' (recovery and residual management) making '5R' as guiding factors to waste management. The adoption of 5-R processes can lead to substantial reduction on the impact of waste on global warming. The reduction of quantities of waste for dumping due to 5-Rs, would lead to less damage to environment.

ii. Recycling of waste for reuse, such as metallic scraps and engineering components after reconditioning shall result in conservation of scarce material and energy resources.

iii. The use of waste as energy sources can supplement the alternate energy needs, esp. the bio-fuels made through utilizing solid (biogas) and liquid waste (sea-weed cultivation).

iv. Successful waste management practices can convert waste into resources, thus reducing the impact of anthropogene on the global warming.

REFERENCES

1. *Global Warming - A Waste*, EPI PDF. file: http://www. epa.gov/ epawaste /partnerships/wastewise/pubs/wwupdate18.pdf).

2. Materials World, British journal, April, 2008.

3. R.Chattopadhyay, invited talk on'Surface Engineering to Control Global Warming', Surface Engineering Seminar, sponsored by Indian Institute of Metals, Tata Research and Development Centre, Pune, 23 January, 2009.

4. *Inventory of U.S. Greenhouse Gas Emissions and Sinks: 1990-2000*, U.S. EPA, Office of Atmospheric Programs, April 2002.EPA236-R-02-003.

5. R.Chattopadhyay, *Surface Wear, Analysis, Treatment and Prevention*, ASM International, USA, 2001.

6. *The Economist*', Jan19th, 2008, advertisement by 'Nickel Institute'.

7. Source: http://www.energyefficienthomearticles.com)

19

Emission Trading and Business

Emission trading is trading of carbon credit permits. Emitters that need to increase their emission allowance can buy credits for the offset amount from those who pollute less or savers. The transfer of set allowances through buying and selling is referred to as a trade. Thus those who can reduce emissions below the allowed limit are financially benefited by selling the carbon units to the defaulters. Emission trading or its popular version 'cap-and-trade' is similar to a tax, *i.e.*, for every cap placed, there exists a tax that has identical economic effects.

Business transactions on emission allowances are carried out in between an emitter and saver or through open market. Markets can be mandatory (for Kyoto Protocol signatory countries, *e.g.*, EU) and voluntary markets (for non-signatories of Kyoto Protocol, *e.g.*, USA). Transaction mechanisms are covered in Kyoto Protocol. The emission trading market is vast. List of emitters include fossil fuel based power plants, transportation, cement & steel industries. The savers list include green constructions, forest preservation, and power generation using alternative energy sources, such as solar, wind and hydro. Business for the savers is growing fast thus providing an opportunity to control the emission rate despite the ever increasing presence of polluters. In addition to industrial sectors, there has been a spurt of green activities in different fronts, such as, real estate, green cities and in service sector.

Emission Trading or Cap and Trade of Emissions

The emission trading, also known as cap and trade, basically refers to the trading of emissions allowances (carbon credit permits), as follows:

(i) between one regulated entity and a less pollutive entity;

(ii) through an intermediate company that sells carbon credits to commercial and individual customer to reduce the *carbon footprint* on a voluntary basis;

(iii) *carbon offsetters* purchasing the credits from an investment fund or a carbon development company that has aggregated the credits from individual projects; and

(iv) from open market.

There are two types of markets for emission trading, as follows:

(i) Mandatory market: for those trading on mandatory carbon offsets as required by Kyoto Protocol; and

(ii) Voluntary market: which provides various options for off setters outside the Kyoto regime.

Kyoto Protocol and Emission Trading[1]

Emissions trading, as set out in Article 17 of the Kyoto Protocol, allows countries that have emission units to spare (*i.e.* units saved through less emission than the permitted limit) to sell the excess units to countries that are over their targets. Thus, a new commodity has been created in the form of emission reductions or removals. Since the global warming potential (GWP) is expressed in terms of carbon dioxide equivalent, the emission unit is simply referred to as *carbon unit* and the trading as *carbon trading*. Carbon is now tracked and traded like any other commodity. The big carbon market across the globe has grown several folds.

Emission certificates/units for trading

More than actual emissions units can be traded and sold under the Kyoto Protocol's emissions trading scheme. The units which may be transferred under the scheme, each equal to one tonne of CO_2, may be in the form of:

- Certified Emission Reduction (CER): a certified emission reduction (CER) generated from a clean development mechanism project activity
- Emission Reduction Unit (ERU): an emission reduction unit (ERU) generated by a joint implementation project
- A removal unit (RMU) on the basis of land use, landuse change and forestry (LULUCF) activities such as reforestation
- Verified Emission Reduction (VER).
- EU has a unified certified emission management and reduction scheme, known as, Certified Emissions Measurement and Reduction Scheme (CEMERS).

In order to do trading on carbon credits, either in mandatory or in voluntary markets it is essential to get the carbon reduction units duly certified and validated by appropriate authorities. Although both markets trade with same emission certificates, but the certificates provided in voluntary markets are not registered by the governments of the host countries or the Executive Board of the UNO.

Tranfer and acquisition registries and International transaction log

Emission targets for industrialized country (Parties to the Kyoto Protocol) are expressed as levels of allowed emissions, or "assigned amounts", over the 2008-2012 commitment periods. Such assigned amounts are denominated in tonnes (of CO_2 equivalent emissions) known informally as "Kyoto units".

The ability of Parties to add to their holdings of Kyoto units (*e.g.* through credits for CDM or LULUCF activities) or move units from one country to another (*e.g.* through emissions trading or JI projects) requires registry systems for tracking and recording of the transfers and acquisitions of the Kyoto units.

Two types of registry are being implemented

Governments of the 38 Annex B Parties are implementing national registries, containing accounts within which units are held in the name of the government or in the name of legal entities authorized by the government to hold and trade units.

The UNFCCC secretariat, under the authority of the CDM Executive Board, has implemented the CDM registry for issuing CDM credits and distributing them to national registries. Accounts in the CDM registry are held only by CDM project participants, as the registry does not accept emissions trading between accounts.

In addition to recording the holdings of Kyoto units, these registries "settle" emissions trades by delivering units from the accounts of sellers to those of buyers, thus forming the backbone infrastructure for the carbon market

Registry operation mechanism

Each registry will operate through a link established with the *international transaction log* put in place and administered by the UNFCCC secretariat. The ITL verifies registry transactions, in real time, to ensure they are consistent with rules agreed under the Kyoto Protocol. The ITL requires registries to terminate transactions they propose that are found to infringe upon the Kyoto rules. An international transaction log ensures secure transfer of emission reduction units between countries. In order to address the concern that Parties could "oversell" units, and subsequently be unable to meet their own emissions targets, each Party is required to hold a minimum level of ERUs, CERs, AAUs and RMUs in its national registry. This is known as the *"commitment period reserve"*

The quality of the credits depends on the validation process and sophistication of the fund or development company that acted as the sponsor to the carbon project. This is reflected in their price; voluntary units typically have less value than the units sold through the rigorously-validated *Clean Development Mechanism*.

Certified Emission Reductions (CERs)

These are issued in accordance with stipulations in the Clean Development Mechanism (CDM). Emission reductions achieved by CDM projects are to be verified by a DOE under the rules of the Kyoto Protocol.

CERs can be used by (a) Annex 1 countries in order to comply with their emission limitation targets or (b) By operators of installations covered by the European Union Emission Trading Scheme (EU ETS) in order to comply with their obligations to surrender EU Allowances of CERs or (c) Emission Reduction Units (ERUs) for the CO_2 emissions of their installations

CERs can be held by governmental and private entities on electronic accounts. CERs are split into long-term (lCER) or temporary (tCER), depending on the likely duration of their benefit. At present, most of the approved CERs are recorded in CDM Registry accounts only. It is only when the CER is actually sitting in an operator's trading account that its value can be monetized through being traded. The UNFCCC's International Transaction Log has already validated and transferred CERs into the accounts of some national climate register. European operators are waiting for the European Commission to facilitate the transfer of their units into the registries of their Member States. Under CDM, several ultra mega thermal power projects are in the process of installation in India (chapter 12).

Emission Reduction Unit (ERU)

The ERU refers to the reduction of GHGs particularly under Joint Implementation, where one unit represents one tonne of CO_2 equivalent reduced. Regarding emission reduction units, one example is the production of biogas by landfill sites. These gases consist of mainly methane which escapes to the atmosphere if it is not collected. The main reason for dealing with methane is that its global warming potential (GWP) has a multiplier of 21 compared to carbon dioxide. Collection of methane is usually accompanied by its combustion. Although burning methane produces carbon dioxide, its greenhouse effect is reduced by 18 ERU. *One tonne of methane, with 21 tonnes carbon dioxide equivalent produces nearly 3 tonnes of carbon dioxide on burning, thus making a net saving of 18 tons of carbon dioxide or ERU.*

Verified Emission Reductions (VERs)

A unit of greenhouse gas emission reduction that has been verified by an independent auditor, but that has not yet undergone the procedures and may not yet have met the requirements for verification, certification and issuance of CERs (in the case of the CDM) or ERUs (in the case of JI) under the Kyoto Protocol. Buyers of VERs assume all carbon-specific policy and regulatory risks (*i.e.* the risk that the VERs are not ultimately registered as CERs or ERUs). Buyers therefore tend to pay a discounted price for VERs, which takes the inherent regulatory risks into account. VER cannot be used in mandatory markets.

Mandatory Market

To reach the goals defined in the Kyoto Protocol with least economical costs the following flexible mechanisms were introduced for the mandatory market:

- Emissions trading
- Clean Development (CDM)
- Joint Implementation (JI)

Emission Trading

The Kyoto Protocol defines legally binding targets and timetables for cutting the greenhouse-gas emissions of *industrialized countries* (excepting USA, not ratified KP) that ratified the Kyoto Protocol. Under the treaty, for the 5-year compliance period from 2008 until 2012, nations that emit less than their quota will be able to sell emission credits to others that exceed their quota. Emission quotas were agreed by each participating country, with the intention of reducing their overall emissions by 5.0% of their 1990 levels by the end of 2012.

For trading purposes, one allowance or CER is considered equivalent to one metric tonne of CO_2 emissions. These allowances can be sold privately or in the international market at the prevailing market price. Also allowances can be transferred between countries. Each international transfer is validated by the UNFCCC. Each transfer of ownership within the European Union is additionally validated by the European Commission.

Clean Development and Joint Implementation

It is also possible for developed countries within the *trading scheme to sponsor carbon projects* that provide a reduction in greenhouse gas emissions in other countries, as a way of generating tradable carbon credits. The Protocol allows this scheme through Clean Development Mechanism (CDM) and Joint Implementation (JI) projects, in order to provide flexible mechanisms to aid regulated entities in meeting their compliance with their caps. The UNFCCC validates all CDM projects to ensure they create genuine additional savings and that there is no leakage.

Voluntary Market

In contrast to the strict rules set out for the mandatory market, the voluntary market provides companies with different options to acquire emissions reductions. An emission reduction scheme, comparable with those developed for the mandatory market, has been developed for the voluntary market, and known as the Verified Emission Reductions (VER). This measure has the great advantage that the projects/activities are managed according to the quality standards set out for CDM/JI projects but the certificates provided are not registered by the governments of the host countries or the Executive Board of the UNO. As such, high quality VERs can be acquired at lower costs for the same project quality. However, at present VERs can not be used in the mandatory market.

Voluntary market in USA

USA not being a signatory to Kyoto Protocol has voluntary markets for emission trading. In 2003, U.S. corporations were able to trade CO_2 emission allowances on the Chicago Climate Exchange under a voluntary scheme. The voluntary market in North America is divided between (i) Chicago Climate Exchange and (ii) Over the Counter (OTC) market. The Chicago Climate Exchange is a voluntary yet legally binding cap-and-trade emission scheme, whereby members commit to the capped emission reductions and must purchase allowances from other members or offset excess emissions. The OTC market does not involve a legally binding scheme and a wide array of buyers from the public and private spheres, as well as special events that want to go carbon neutral.

In 2007, the California passed the California Global Warming Solutions Act, AB-32, providing flexible mechanisms in the form of project based offsets for five main project types. A carbon project would create offsets by showing that it has reduced carbon dioxide and equivalent gases. The project types include: manure management, forestry, building energy, SF6, and landfill gas capture. California is also one of seven states and three Canadian provinces that have joined together to create the Western Climate Initiative, which has recommended the creation of a *regional greenhouse gas control and offset trading environment.*

There are project developers, wholesalers, brokers, and retailers, as well as carbon funds, in the voluntary market. Some businesses and nonprofits in the voluntary market encompass more than just one of the activities listed above. A report by Ecosystem Marketplace shows that *carbon offset prices increase as it moves along the supply chain i.e., from project developer to retailer.*

While some *mandatory emission reduction schemes exclude forest projects, these projects flourish in the voluntary markets.* A major criticism concerns the imprecise nature of GHG sequestration quantification methodologies for forestry projects. However, others note the community co-benefits that forestry projects foster. Project types in the *voluntary market include avoided deforestation, afforestation/reforestation, industrial gas sequestration, increased energy efficiency, fuel switching, methane capture from coal plants and livestock, and even renewable energy.*

Renewable Energy Certificates

Renewable energy certificates (RECs), also known as green certificates, green tags, or tradable renewable certificates, represent the environmental attributes of the power produced from renewable energy projects. RECs can be traded through retailers, wholesalers and certificate brokers/exchanges. Several agencies for consumer protection and the REC tracking systems are also available. Renewable Energy Certificates (RECs) sold on the voluntary market are quite controversial due to additional concerns. Industrial Gas projects receive criticism because such projects only apply to large industrial plants that already have high fixed costs.

The size and activity of the voluntary carbon market is difficult to measure but projected to be $4 billion by 2010.

Credits versus Taxes

Credits were chosen by the signatories to the Kyoto Protocol as an alternative to *Carbon taxes*. A criticism of tax-raising schemes is that they are frequently not *hypothecated*, and so some or all of the taxation raised by a government may be applied inefficiently or not used to benefit the environment. By treating emissions as a market commodity it becomes easier for business to understand and manage their activities, while economists and traders can attempt to predict future pricing using well understood market theories. Thus the main advantages of a tradable carbon credit over a carbon tax are:

- fair price of carbon, as set by the market, and investors in credits have more control over their own costs.
- investment goes into genuine sustainable carbon reduction schemes through accepted validation process in the flexible mechanisms of Kyoto Protocol.

Market price for carbon

The energy use and hence emission levels are going to increase with time. With increasing demand and less credit availability would push up the market price following the rules of supply and demand. Higher price of credits shall encourage more groups to initiate measures to curb emissions which would create carbon credits to sell.

An individual allowance, such as a Kyoto Allocation Allowance Unit (AAU) or its near-equivalent European Union Allowance (EUA), may have a different market value to an offset such as a CER. The offsets generated by a carbon project under the Clean Development

Mechanism are potentially limited in value because operators in the EU- ETS are restricted as to what percentage of their allowance can be met through these flexible mechanisms.

Market Trend

Climate exchanges have been established to provide a *spot market* in allowances, as well as *futures* and *options market* to help discover *a market price* and maintain *liquidity*. Carbon prices are normally quoted in European market as Euros per tonne of carbon dioxide or its equivalent (CO_2e). Other greenhouse gasses can also be traded, but are quoted as standard multiples of carbon dioxide with respect to their global warming potential (GWP). These features reduce the quota's financial impact on business, while ensuring that the quotas are met at a national and international level. Currently there are at least four exchanges trading in carbon allowances, such as, Chicago Climate Exchange, European Climate Exchange, Nord Pool, and Power Next.

The last one has a contract to trade offsets generated by a CDM *carbon project* called Certified Emission Reductions (CERs). Many companies engaged in emissions abatement, offsetting, and sequestration programs to generate credits operate through these exchanges.

Market Size

Managing emissions is one of the fastest-growing segments in financial services, especially in the City of London. With the creation of a *market for mandatory trading* of carbon dioxide emissions within the Kyoto Protocol, the *London financial marketplace* has established itself as the center of the carbon finance market. The market is now worth about $60 billion (€30 billion) in 2007, expected to grow into a market of $0.5 trillion (€1 trillion) within a decade.

The voluntary offset market, by comparison, is projected to grow to about $4bn by 2010. Louis Redshaw, head of environmental markets at Barclays Capital predicts that "Carbon will be the world's biggest commodity market, and it could become the world's biggest market overall"

The quantum of Carbon emissions trading has been steadily increasing in recent years. According to the World Bank's Carbon Finance Unit, 374 million metric tonnes of carbon dioxide equivalent (tCO_2e) were exchanged through projects in 2005, a 240% increase relative to 2004 (110 $mtCO_2e$) which was itself a 41% increase relative to 2003 (78 $mtCO_2e$).

In terms of dollars, the World Bank has estimated that the size of the carbon market was 11 billion USD in 2005, 30 billion USD in 2006, and 64 billion in 2007. In near future, the price of carbon will be the most pressing question.

In America, the new president has pledged to cut emissions by instituting a cap-and-trade scheme. EU is fine tuning the next phase of carbon trading scheme. Next updating of Kyoto Protocol by 2012, would decide the modalities of carbon trading practices, based on global accord.[3]

Business Reaction[3]

Twenty three multinational corporations came together in the G8 Climate Change Roundtable, a business group formed at the January 2005 World Economic Forum. The group included Ford, Toyota, British Airways, BP and Uniliver. The group has published a statement stating that there was a need to act on climate change and stressing the importance of market-based solutions. It called on governments to establish "clear, transparent, and consistent price signals" through "creation of a long-term policy framework" that would include all major producers of greenhouse gases. By December 2007 this had grown to encompass 150 global businesses.

Businesses in the UK have come out strongly in support of emissions trading as a key tool to mitigate climate change, supported by Green NGOs.

Other Activities

Carbon Footprints[3]

The *carbon footprint* is a measure of the exclusive global amount of GHGs greenhouse gases, emitted by a human activity or accumulated over the full life cycle of a product or service and expressed as a carbon dioxide equivalent (usually in kilograms or tonnes). The carbon footprints can be assessed for a product, corporate or an event (like X'mas when carbon footprint found to increase substantially in UK).

A carbon label, which shows the carbon footprint embodied in a product in bringing it to the shelf, was introduced in the UK in March 2007 by the Carbon Trust. Examples of products featuring their carbon footprint are Walkers Crisps, Innocent Drinks, and Boots shampoos. A conceptual tool in response to carbon footprints are carbon offsets, or the mitigation of carbon emissions through the development of alternative projects such as solar or wind energy or reforestation.

Carbon Trust[3]

The Carbon Trust is non - profit company(trust) to help businesses and public organizations to reduce their emissions, by funding development of low carbon technology, such as, development of alternate renewable energy. Unusual for a government sponsored organization it operates venture capital and loan funds as well. It is partly funded from the Climate Change Levy, a tax on electricity, gas, and coal. The Trust's annual report for 2008 indicates an estimated savings for UK business of £1 million a day through the cost savings by reducing carbon emissions.

Carbon Banking[3]

Several forms of "carbon banking" accounting are being developed. Carbon banking is a service that itemizes carbon reductions and storages, and provides some form of economic compensation as the global warming issue heats up. Some of the carbon banking is as follows:

GAIA Carbon Bank, trying to purchase existing forests in the northern hemisphere and to manage these as a future ecologically sounds resource. *Carbon Bank USA*, in their environmental outreach program, provides an opportunity to reduce carbon emissions by planting trees and reforestation.

Morgan Stanley has recently announced the creation of the *Morgan Stanley Carbon Bank* to assist clients seeking to become carbon neutral. Offsetting capabilities based on the highest recognized international standards. Britain has recently urged for the creation of an independent European carbon bank to improve the functioning of the EU's.

The World Bank Carbon Finance Unit's (CFU) are part of the larger global effort to combat climate change, along with the World Bank and its Environment Department. The CFU uses money contributed by governments and companies in OECD countries to purchase project-based greenhouse gas emission reductions in developing countries and countries with economies in transition. The emission reductions are purchased through one of the CFU's carbon funds on behalf of the contributor, and within the framework of the Kyoto Protocol's Clean Development Mechanism (CDM) or Joint Implementation (JI).

Unlike other World Bank development products, the CFU does not lend or grant resources to projects, but rather contracts to purchase emission reductions similar to a commercial transaction, paying for them annually or periodically once they have been verified by a third party auditor. The selling of emission reductions - or carbon finance - has been shown to increase the bankability of projects, by adding an additional revenue stream in hard currency, which reduces the risks of commercial lending or grant finance. Thus, carbon finance provides a means of leveraging new private and public investment into projects that reduce greenhouse gas emissions, thereby mitigating climate change while contributing to sustainable development.

Emission Trading Scheme to help farmers.

In USA, Farmers Union's Carbon Credit Program allows agriculture producers and landowners to earn income by storing carbon in their soil through non-till crop production and long-term grass seeding practice. For example, the Farmers Union has earned approval from Exchange, like Chicago Exchange, to aggregate carbon credits. Farmers Union will enroll producers' acreage of carbon into blocks of credits that will be traded in the Exchange, much like other agricultural commodities that are traded.

Green buildings

California has adopted first statewide green building code. Approved by state's Building Standards Commission in July, 2008, the new standards are designed to encourage construction of new buildings to use at least 15% less energy, conserve water, and reduce the use of products containing toxic substances. Building more energy efficient buildings and homes is the state's strategy to comply with its mandate to cut greenhouse gases to 20% below 2005 levels by 2020, or by about 30%. Initially voluntary, the new standards become mandatory with the state's 2010 According to building code, California may only credit efforts that produce GHG reductions that are real, quantifiable, permanent, verifiable, enforceable and additional beyond those that would otherwise occur under business as usual. The same general requirements govern offsets in other GHG programs throughout the world. These are the threshold design issues a Green Building Credit must satisfy. In some systems, a customized process is used for each project, such as in the Clean Development Mechanism under the Kyoto Protocol. But such project-by-project analysis creates high transaction costs that undermine the financial viability of offset projects. The preferred and more effective method is to use standardized measurement and verification protocols[4].

Green cities[3]

Based on the (i) *source of electricity* such as wind, solar, biomass and hydroelectric power, residents using their own power sources, like roof-mounted solar panels, (ii) *type of transportation* such as public transportation or carpool (iii) *green living* such as number of buildings certified by the U.S. Green Building Council, as well as for devoting area to *green space*, such as public parks and nature preserves and (iv) *recycling and green perspective* based on a measure of how comprehensive is city's recycling programme and citizen's participations, 50 cities in USA have been rated as green cities.

Green Sports[3]

NBA's New Jersey Nets has bought 'carbon credits' that support 4-hydro-power stations in China, as part of programme to offset its own carbon emissions in America. The Nets claim to be first major professional sports team to achieve carbon neutral status[3].

Indian Scenario in Carbon Trading[5]

The government is betting big on two market-based trading schemes to encourage energy efficiency and green power across the country sidestepping widely used cap & trading emissions schemes. The trading scheme is called as Perform, Achieve and Trade (PAT). India is starting a mandatory scheme that sets benchmark efficiency levels for 563 big polluting from power plants to steel mills and cement plants that account for 54 percent of the country's energy consumption. The scheme allows businesses using more energy than stipulated to buy tradable energy saving certificates, or Escerts, from those using less energy, creating a market estimated by the government to be worth about $16 billion in 2014 when trading starts. The number of Escerts depends on the amount of energy saved in a target year

Summary

1. Emission trading is trading of carbon credit permits, which an emission offsetter can buy from those who pollute less or savers. The transfer of allowances is referred to as a trade.
2. Emission trading can be between two parties or through open market. Markets can be mandatory for Kyoto Protocol signatory countries, (*e.g.*, EU) and voluntary for others.
3. Transaction mechanisms for mandatory are covered in Kyoto Protocol. Each unit is equal to one tonne of CO_2, may be in the form of CER from clean development mechanism, ERU from joint implementation, RMU from land use and forestry or VER. Voluntary market trades on VER and also REC. Market is big and covered by various exchanges and trading agencies. Same units are also used

in voluntary market. The World Bank Carbon Finance Units (CFU) are used to help less developed nations in implementing emission saving green projects.

REFERENCES

1. Kyoto Potocol Documents.
2. Welding Journal, American Welding Society, Oct, 2008, p4—'Washington Watchdog', by Waugh K. Webster.
3. The Economist, special issue, The World in 2009, *The Environment*, p. 105, 106.111
4. Support for Green Building Carbon Credits - BOMA, U.S Green Building Council, http://ghgblog.com/?p=672
5. India takes unique path to lower carbon emissions, Reuters report, Times of India, 31[t] May, 2011.

20

Corporate Social Responsibility and Climate Change

Corporate Social Responsibility (CSR), also called 'corporate responsibility', or 'corporate citizenship', is the responsibility of the business house to 'building sustainable businesses' without forgetting their social obligations including protection of the environment.

Industries are mainly responsible for anthropogenic emissions causing global warming and the consequent climate change. In addition to GHG emissions, the industrialization is linked to deforestation, pollution of water resources, destruction of biome & biodiversity thus reducing the sinks for carbon dioxide. However the industries are here to stay and also need to grow at an accelerated rate to cater to the growing needs of people across the globe. In the preceding chapters, the direct or indirect emission management processes in various sectors of the industries are described. It is imperative that corporates should shoulder the responsibilities of limiting emissions, at least to an extent specified in Kyoto Protocol, if not more.

Major part of emission is from developed countries. While being not major emitters, developing nations are of major concern in the area of deforestation and destruction of biodiversity, the main sinks for GHGs. The developed countries, which are signatories to

Kyoto Protocol, have the mandatory emission reduction targets for their industries. The CSR activities with respect to climate change are mandatory for industries in Kyoto Protocol signatory nations. The non-signatories to Kyoto Protocol, such as USA, Australia, and Canada have voluntary carbon trading market providing financial incentives to the carbon saver industries.

However beyond emission reductions in their respective corporations, the corporate need to protect and care for preserving natural resources, such as, soil, water, forest, biodiversity, which act as natural buffers for GHGs, in the nature's carbon and other cycles. CSR activities can play an important role in saving the globe from disasters associated with climate change.

Defining Corporate Social Responsibility

Most definitions of CSR emphasize the interrelationship between economic, environmental and social aspects and impacts of an organization's activities on societies in general. Social responsibility "*is taken to mean a balanced approach for organizations to address economic, social and environmental issues in a way that aims to benefit people, communities and society.*"

In a nutshell CSR deals[1] with the 3 dimensions - 'economy', 'ecology' and 'social' in the frame of a corporate viewpoint, which might be an accumulation of different opinions (from the employees of that company), and/or on a specific engagement (of one of the founders, managers, etc.)

Corporate Social Responsibility is a concept whereby companies integrate social, environmental and financial concerns into the business operations and in their interaction with their stakeholders (employees, customers, shareholders, investors, local communities, government), on a voluntary basis in order to enhance sustainability and growth (tab.19.1). CSR is closely linked with the principles of sustainability, which argues that enterprises should make decisions based not only on financial factors such as profits or dividends, but also based on the immediate and long-term social and environmental consequences of their activities.

Table 19.1 CSR Activities

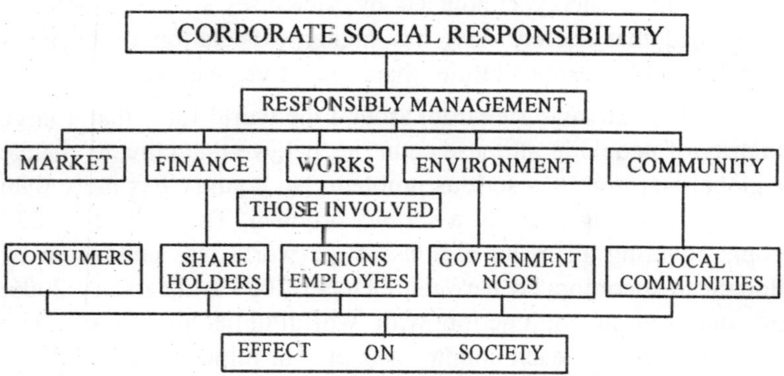

Society in general is affected by mismanagement in any one of the five areas, such as, market, finance, works, environment and community (tab.19.1). Earlier business practice was confined to three business-related areas, viz., market, finance and works. There has been a paradigm shift in corporate management activities from earlier business practices with the introduction of CSR in a big way in all major corporations. The environment and community constitute major part of CSR. *The most challenging concurrent environment related problem is global warming –a problem created and sustained mostly due to contributions from corporations and therefore it is the responsibility of the corporates to work towards a techno-ecologically viable solution to limit global warming.*

Growth of CSR

CSR, once a do-something sideshow, is now not only in mainstream, but booming. In recent times, the corporate giants are telling the world of their CSR activities through electronic and print media, in their advertising campaigns, in glossy reports and posters. CEOs queue up to speak at conferences about their love for the community or their new found commitment to make their company carbon neutral. A recent survey shows corporate responsibility rising sharply in global executives' priorities[2].

The biggest single driver of growth in corporate social responsibility (CSR) is ascribed as due to concern over climate change. Al Gore's initiative[2a] to propel climate to the top of global agenda has led to great green awakening. The leading companies and their CEOs are having a serious look on the impact on environment due to their activities. In a survey conducted by McKinsey in 2007, 95% of CEOs said that society now has higher expectation of business taking on public responsibilities than it had five years ago.[2]

A poll of 30 countries from around the world finds that a large majority of people in all these countries believe that climate change or global warming is a serious problem. No country has more than one in five saying it is not a serious problem. The poll of 33,237 people covering all major regions of the world was conducted by GlobeScan Incorporated between October 2005 and January 2006, and analyzed in conjunction with WorldPublicOpinion.org. The margin of error for each country was plus or minus 3 percent.

Perhaps most significant, in 23 countries a majority says that global warming is a "very serious" problem. On average, 65 percent say that it is a very serious problem. The only countries where this is not a majority position are six developing countries (China 39%, Indonesia 44%, Kenya 44%, South Africa 44%, Philippines 46%, Nigeria 47%,) and the US (49%).

Current trend for major companies is to produce their annual report on CSR including energy savings, GHG emissions, carbon footprint etc. CSR is rapidly becoming a major part of all business management courses and a key global issue.

CSR Global Issue

MNCs representing big corporates need to formulate their CSR activities to cater for the needs and aspirations of local communities, and also to cover the stipulations made in rules and regulations of the country. Today's world, any irresponsible act done in any part of the globe finds it way by communication super-highway in no time to print and electronic media which flashes the news in their prints and 24x7 broadcast. Thus global CSR has become prominent in the language and strategy of business activities for MNCs.

Guidelines for CSR

OECD

The Organization for Economic Co-operation and Development (OECD) *defines corporate codes* as "commitments voluntarily made by companies, associations, or other entities, which put forth standards and principles for the conduct of business activities in the marketplace".

The organization has provided OECD -2000 Guidelines on CSR for its member countries. The guidelines have the government support but are voluntary CSR activities. Member of the OECD include most of European nations, Canada, United States, Japan, Australia, New Zealand, Mexico and Korea.

The guidelines on CSR activities are prepared by OECD for the multinational enterprises to make positive contributions to economic, environmental and social progress and to minimize the difficulties to which their various operations may give rise. In working towards this goal, governments find themselves in partnership with the many businesses, trade unions and other non-governmental organizations that are working in their own ways toward the same end. The voluntary guidelines for CSR with respect to environment are provided in the UN global compact and also in the ISO-14000-2004.

UN Global Compact

The dedicated CSR organizations, Governments and international organizations (UN) are increasingly encouraging CSR[3,4,5].

Launched in 2000, the UN Global Compact brings business together with UN agencies, labour, civil society and governments to advance ten universal principles in the areas of human rights, labour, environment and anti-corruption. It is a strategic policy initiative for businesses that are committed to aligning their operations and strategies with *ten universally accepted principles* in the areas of *human rights, labour, environment and anti-corruption.* By doing so, business, as a primary agent driving globalization, can help ensure that markets, commerce, technology and finance advance in ways that benefit economies and societies everywhere.

The Global Compact is a leadership platform, endorsed by CEOs, and offering a unique strategic platform for participants to advance their commitments to sustainability and corporate citizenship. Structured as a public-private initiative, the Global Compact is policy framework for the development, implementation, and disclosure of sustainability principles and practices. It offers the participants a wide spectrum of specialized workstreams, management tools and resources, and topical programs and projects - all designed to help advance sustainable business models and markets in order to contribute to the initiative's overarching mission of helping to build a more sustainable and inclusive global economy[3].

The three environmental principles of the Global Compact are drawn from the 10 principles of Rio Earth Summit, in 1992, (Chapter 9). These three principles are as follows:

1. Principle (no.7 of Rio) - business should support a precautionary approach to environmental challenges
2. Principle (no.8 of Rio) - undertake initiatives to promote greater environmental responsibility
3. Principle (no.9 of Rio) - encourage the development and diffusion of environmentally friendly technologies

The environmental principles of the Global Compact provide an entry point for business to address the key environmental challenges. In particular, the principles direct activity to areas such as research, innovation, co-operation, education, and self-regulation that can positively address the significant environmental degradation, and damage to the planet's life support systems, brought by human activity.

UN global compact has been criticized for its 'soft codes' that the corporates need to follow and therefore getting quite popular. In order to sign up the companies need to commit themselves to basic principles as outlined above and report on their progress once a year.

The main criticism against the compact is that the toothless character of the compact provides cover for companies from China and elsewhere. They happily sign up and then more happily ignore it[4]. In order to counter the criticism, United Nations Global Compact has implemented a range of measures to promote accountability among companies that participate in the pogramme. They are required to give annual updates on the progress they made in aligning themselves with the initiative's 10 principles[5].

ISO 14000-2004

The ISO 14000 family addresses various aspects of environmental management. The very first two standards, ISO14001:2004 and ISO14004 :2004 deal with environmental management systems (EMS). ISO 14001 :2004 provides the ***requirements*** for an EMS and ISO 14004:2004 gives general EMS ***guidelines.***

The requirements of ISO 14001:2004 is a management tool enabling an organization of any size or type to:

- identify and control the environmental impact of its activities, products or services;
- to improve its environmental performance continually;
- to implement a systematic approach to setting environmental objectives and targets, to achieving these; and
- to demonstrating that they have been achieved.

ISO14000 for environment management has been an accepted practice for a large a large number of business establishments, who have found it beneficial for sustaining the profitable business with less of waste, pollution & risk, but higher productivity.

However, due to voluntary nature of CSR, companies following this or other guidelines, such as, from ILO, OECD, as well as standards, such as, ISO 14001 (for the environment) and SA 8000 (for human rights) or ISO 2600 (for social responsibility) do have a large number of black sheeps in the herd.

A conference sponsored jointly by UN Environmental Programme Finance Initiative (UNEP-FI) and UN-backed Principles for Responsible Investment (PRI) was held in Seoul, in June' 2008. The list of participants include representatives of global financial institutions worth 14 trillion US dollars in assets and leading representatives of Asian business. Basic aims of the conference, amongst others, were (i) to promote better environmental, social and governance performance in business and (ii) calls on conference participants to, inter alia, join forces around efforts to address the carbon challenge. The conference also provides details of what financial institutions in the Republic of Korea and broader Asia can do on climate change, water, biodiversity and social issues. The issues of climate change and corporate governance have converged to provide a solid business case for mainstream financial institutions to apply responsible investment approaches to their core activities.[6]

CSR and Kyoto Protocol

CSR is considered as voluntary activity of the corporate. The three basic principles on environment as incorporated in UN global compact provide the guidelines for the corporates to follow to prevent environmental degradation. While Rio Earth Summit dealt with the voluntary protection of environment, the Kyoto Protocol defines legally binding targets and timetables for cutting the greenhouse-gas emissions of *industrialized countries* (excepting USA, not ratified KP) that ratified the Kyoto Protocol. It is a mandatory requirement of the corporates operating in these countries to reduce emission figure to a quantified limit as stipulated by the regulatory authorities. The net changes in the emission figure in terms of carbon equivalent are a tradable commodity, called 'carbon credit'. The corporation can earn through carbon credits sales by reducing carbon emission for the stipulated limit. On the other hand, if the limit is exceeded, they need to buy carbon credits for extra emission. This is a costly option for the corporation. The defaulting corporations should adopt some of the techno-ecologically viable and sound economic practices. The list of such practices includes the followings:

i. Improvement of energy efficiency in the relevant sectors of industries.

ii. Protection and enhancement of GHG sinks and reservoirs (forest, water, ocean) in compliance with international agreements and national regulations.

iii. Promotion of sustainable forest management practices.

iv. Promotion of sustainable form of agriculture in light of climate change

v. Research on, and promotion, development of renewable energy, CO_2 sequestration, and related advanced environmentally sound technologies.

vi. Progressive reduction or phasing out of market imperfections, fiscal incentives, tax and duty exemptions and subsidies in all greenhouse gas emitting sectors that run counter to the objective of Kyoto Protocol.

vii. Encouragement of appropriate reforms in relevant sectors aimed at promoting policies and measures which limit or reduce emissions of GHGs

viii. Measures to limit and /or reduce emissions of GHGs in the transport.

ix. Limit/reduce methane emissions through recovery and use in waste management.

x. Limit/reduce in GHGs from aviation and marine bunker fuels, through International Civil Aviation Organization and the International Maritime Organization respectively.

The implementation of all these measures in controlling emissions and the results are discussed in preceding chapters. Corporate legal responsibilities end in satisfying the mandatory requirements of the countries under Kyoto Protocol, but not the moral or ethical part of CSR. In USA, corporates are not covered by Kyoto Protocol, hence the CSR with respect to climate change is voluntary.

CSR activities, excepting the mandatory requirements by Kyoto Protocol and other governmental regulations, are in general considered as voluntary. However due to faulted practices by several corporations, there has been a popular demand to make CSR as mandatory.

Potential Benefits for Corporations

No matter the size of an organization or the level of its involvement with CSR, every contribution is important and provides a number of benefits to both the community and business. Contributing to and supporting CSR does not have to be costly or time consuming and more and more businesses active in their local communities are seeing significant benefits from their involvements, including followings:

- Reduced costs
- Increased business leads
- Increased reputation

- Increased staff morale and skills development
- Improved relationships with the local community, partners and clients
- Innovation in processes, products and services
- Managing the risks a company faces
- Boost image of company and brand names
- Promote business with green image

Voluntary Vs. Mandatory CSR

The *common thread* that weaves through the various definitions of "Corporate Social Responsibility" is the *voluntary* nature of the good practices referenced. What makes CSR initiatives "socially responsible" is that they are *not* mandated by governmental or intergovernmental institutions — they are voluntarily pursued. The most celebrated *mechanism in the CSR toolkit* is *the corporate code of conduct*. The Organization for Economic Co-operation and Development (OECD) *defines corporate codes* as "commitments voluntarily made by companies, associations, or other entities, which put forth standards and principles for the conduct of business activities in the marketplace"

More and more studies measuring the *(in) effectiveness of voluntary corporate codes* are being published every day. The *consensus* underlying the divergent findings is that, even if voluntary codes have potential, *they are not currently addressing globalization's externalities in a sustained way.*

In an article in The Guardian[7], Tony Juniper has pointed out the corporate misdeeds, which have led to Britain's non-governmental groups calling for new laws to protect people and the environment from the impacts of big business. Some of the facts against corporate are as follows[7]:

- Of some 60,000 multinationals in the world today, only 3% even bother to do a social and environmental report, let alone take any action to reflect its findings. What about the 97% who don't do a report at all? Even those that say they have embraced CSR as a management tool there is a great deal to be concerned about.

- Shell is a well-known CSR proponent, famously claiming that there is no choice between profits and principles and yet the reality of its social and environmental impacts just won't go away. Misdeeds include following:
  i. Right now (2006), Shell is involved in operations that are likely to hasten the extinction of the last population of western Pacific grey whales.
  ii. It is illegally flaring gas next to communities in Nigeria and is aggressively seeking out the new fossil fuels that will accelerate climate change.
- Tesco, along with many other UK companies, uses palm oil in a wide range of products. A vegetable fat derived from oil palms is used to make these products. Plantations of this crop are being aggressively expanded into areas of pristine rainforests across south-east Asia, including through the last frontier regions of Borneo and Sumatra. The rapid loss of the forests that is directly caused by palm oil expansion is leading to the extinction of many species. The orang utan is one species that will disappear if something isn't done soon to tame this rapacious industry. Palm oil expansion is also causing conflicts with forest people and human rights abuses.

Shift towards Green CSR

While the debates continue whether the CSR would be mandatory or voluntary, there has been a paradigm shift on CSR activities. The big change is not only due to pressure groups of NGO's, social and environment activists, but also due to ubiquitous media particularly electronic one flushing the news & visuals across the globe within minutes of happenings. Moreover CSR is no longer considered as a loosing proposition, but a positive endeavor for prosperity and well-being of both the corporations and the societies in which they belong.

Corporations can only gain by following green theme in CSR, since CSR activities allow the company to followings:

i. Reduce GHG emissions, hence earn carbon credits, a tradable commodity adding to the income. With carbon emissions beyond specified limit, carbon credits for the excess carbon need to be purchased at the existing market price.

ii. Encourage 'responsible' use of resources; cut down the over-consumption, work for conservation of resources, develop renewable and sustainable resources. Efforts in these directions shall reduce cost and improve sustainability. Recycling of metallic and other materials from generated scrap leads to not only the conservation of non-renewable resources, but also improve substantially the sustainability of the resource and cost reduction.

iii. Utilize waste or ensure proper disposal by introducing environmental friendly waste management techniques. Good environment and healthy atmosphere in the neighbourhood and workplace improve productivity.

iv. Reduce energy requirement by using energy-efficient processes, process automation etc, which would reduce cost plus earn carbon credit

v. Improve overall efficiency, leads to higher productivity and lower cost of production

vi. Improve sales figure by marketing real green products/ with green tag.

By going green in CSR, companies can improve earnings, ensure sustainable business and would vastly improve company's ratings in national and global perspective.

Social responsibilities combined with financial gains are doubly satisfying –doing well and doing good — are therefore becoming extremely popular.

On the other hand, not going green the corporation would have tough time to sustain business activities in near future, due to followings:

- no sale of products like car not conforming to emission norms (Europe);
- no sale of agricultural products produced in illegally deforested areas (Brazil);

- no sale of products in USA from CSR defaulting countries if new bill in Senate pass through;
- drop in prices due to shifting from elite market to a market place in a less developed country with relaxed norms; and
- drop in share prices due to less sales and lower margin.

Truly Committed Companies on Voluntary CSR

However corporate doing real good work in this area include some big names, such as, the aluminum major, Aluminum Company of America (ALCOA), (see in Chapter 12) and many more.

Examples of some truly committed companies to voluntary CSR are growing. Few examples are as follows:

1. Murata Manufacturing Colt, Kyoto, Japan, is a multinational company produces specialized electrical and electronic goods. CSR Management has yielded good result in terms of improved productivity with lesser carbon emission figures[8].
   - *Murata Group Environment Policy:* Murata strives to quantitatively ascertain the environmental impact associated with its business activities and analyzes that information to reduce the environmental impact of production activities.
   - *Environmental Management*: Murata has obtained ISO14001 certification for the Group's domestic sites and overseas production plants. With this certification, they are moving ahead with environmental management through environmental audits and education and reorganization of the management system.
   - *Environmental Action Plan and Performance*: Murata has formulated its strategy for reducing its impact on the environment in stages in its 4th Environmental Action Plan, with 2010 as the target year. The Group carries out annual action plans to facilitate achievement of the 2010 goals.
   - *Environmental Accounting*: Murata endeavors to reduce the environmental impact of its business activities by

determining and analyzing the cost of environmental protection as well as the effectiveness of the results of its efforts on the environment.

- *Environmentally Conscious Design*: Murata manufactures products with reduced environmental impact throughout their life cycle, including design, production, use, disposal, and recycling.
- *Green Procurement and Purchasing*: Green procurement of part materials with low environmental impact is essential for decreasing the environmental impact of products. That is why Murata asks its suppliers for their understanding and cooperation regarding environmental management.
- *Prevention of Global Warming*: The reduction of greenhouse gases is becoming increasingly urgent. Murata places priority on the reduction of total emissions and per unit of net production emissions of greenhouse gases.
- *Reduction of Waste:* The Murata Group achieved its goal of zero emissions (zero landfill waste and a 100% recycling rate) nationwide in 2003. They are now working toward zero emissions and a reduction in the total amount of waste internationally.
- *Managing Chemical Substances and Environmental Risk*: Murata handles chemical substances, & thus considers their responsibility to prevent environmental pollution. The management of hazardous chemical substances emitted during the production process and the reduction of emissions is an important issues, and they are working to resolve them.
- *Promoting Eco-Friendly Distribution and Packaging*: Murata is reducing its environmental impact during not only production stages but also the product distribution stage. The company is reducing CO_2 by making transport more efficient, and waste reduction by cutting down on packaging materials.

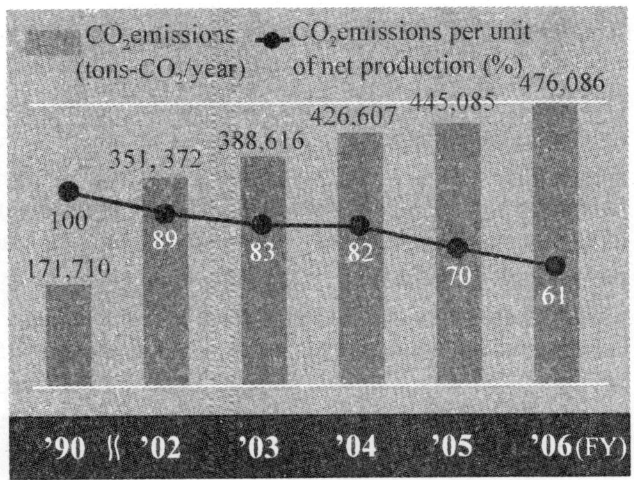

Fig.19.1. Murata's CSR –Green Ratings

Comparison per Unit of Net Production: The JEITA common index, (a measuring standard from JEITA - stands for Japan Electronic Information Technology Association) which is the value of CO_2 emissions against net production (monetary value), adjusted using the corporate goods price index announced by the Bank of Japan. A fall in this figure indicates that a given product quantity (monetary value) is produced using less energy.

Murata is striving to reduce CO_2 emissions by domestic manufacturing plants and subsidiaries by 28% per unit of net production compared with fiscal 1990 levels. By fiscal 2006, the company has slashed this figure by 38.8%., more than around 11% than the set goal of 28%. (Fig.19.1) The net growth over a period of 5 years is 278% in terms of monetary value of the net production. Even with higher energy consumption from the future expansion of plant facilities, with their continued CSR, they expect to meet the reduced energy target for fiscal 2010[9].

Murata's actual emissions were up 5.5% during fiscal 2006 as a result of production expansion. However, they are aggressively advancing with their energy reduction strategies through the introduction of such high-efficiency equipment as double-bundle turbo refrigerators and other measures. During fiscal 2007, they aim to extend energy conservation initiatives from Japan to overseas operations.

2. ***Asahi Glass Co.Ltd, Japan:*** AGC Ltd., manufacture products using materials that meet their green procurement criteria, and deliver environmentally friendly products to customers. The AGC Group uses different types of materials in its four business segments (glass, electronics and displays, chemicals, and others).

   In conformity with the guidelines, AGC obtained work clothes made of blended materials. This means that 55% of the polyester mixed in the fabric will become recycled PET resins. "Green" work clothes have the same color, feel and antistatic performance as previous ones. Against the target green purchasing rate of 98% set for the total purchase of four items, the actual result has been 98% since fiscal year 2005, thus achieving the target.

   In order to increase its competitive edge, the AGC Group holds information exchange meetings, seminars, training sessions, and plant tours in cooperation with its suppliers.

In addition to above examples, brief description of achievements by some truly committed companies[10] are as follows:

3. ***Recreational Equipment Inc (REI):*** America's biggest consumer co-operative. Target is to achieve 'carbon neutral' by 2020, 1/3rd reduction by 2009, from the base level in 2006. Their findings and acts to mitigate carbon emissions:

   - A quarter of carbon emissions came from adventure flights they organized, so they started to buy carbon offsets of these trips.
   - One-fifth of emission results from electricity consumption so they started buying electricity generated by renewable sources, such as hydro power in Washington State.
   - To cut down energy wastage in transport, they opened second distribution hub in Pennsylvania.

- To reduce GHG emissions by employee commuting, which accounts for 1/5th of total emissions, they provide incentives to those who cycle to work.
- They are working on carbon footprints of buildings, papers and packaging and also their products. Green labeling would follow.

4. ***TNT- a Dutch logistics company:*** Peter Baker, heading the company has launched "Planet Me" campaign with the objective of changing company's carbon trajectory. TNT's carbon footprints have been measured and targets for reducing the same would be set. Meanwhile the facts and figure on emissions are as follows:
   - Travel budget has been cut by 20%, (a saving of 3.2 million Euros, compared to cost of installing video-conferencing as 2.8 million Euros) which would cut down emission figure considerably.
   - In 2020, plan to move headquarters to a carbon-positive building.
   - As a global transport company owning a big fleet of trucks and aeroplanes, in 2006, TNT produced 826 killotonnes of carbon dioxide. To cut down emissions from trucks, they are introducing hybrid and electric vehicles. For 44 aircrafts accounting for half of TNT's emissions, they plan to run more efficiently and willing to invest in promising aircraft technologies.
   - Reporting emissions would follow the same rule as financial reporting, including warnings of poor performance like profit warnings, and bonus scheme linked to emissions.

5. **General Electric:** Producers of wide range of products on energy sectors, from electric bulbs to gas turbines for power generation and jet-engines. In 2005 it launched 'ecomagination'-a push to invest in green technology and expand sales of products and services with measurable superior environmental performance on sustainability by at least 10%.

Target: To cut down 1% of GHG emission by 2012 compared to 2004 as baseline.

- GE has already reduced emissions by 4% (2008)
- GE to invest $1.5 billion in 2010 ($700m now) in R&D for cleaner technology
- Expects revenue from ecomagination products of at least $20 billion, GE have sold out all eco-certified products to 2009. For example it is not possible to buy GE's wind turbine before 2010.
- Employees' initiatives have resulted in the savings of $70 m in energy each year.
- GE has joined other big companies and NGOs to form US Climate Action Partnership to lobby for national legislation in America to cap carbon emissions.

6. ***United Technology Corporation (UTC, USA):*** Products range from aerospace to air-conditioners and has been able to *reduce carbon footprint by 19% over the last 10 years, even it has doubled its output*. George David, CEO says' we had an explosion of ***doing more with less***". In 2008, UTC is aiming at 10% growth rate, while cutting carbon emission by another 5%.

7. ***Arcelor Mittal. –Supplier Engagement Program:*** An example of the best practice example in supply chain responsibility includes the one by Arcelor Mittal, a global leading steel company. The company has participated in a supplier engagement program in Brazil with a focus on sustainability and environmental issues and presented lately first results. Of the participating suppliers, 73% have started engaging in waste management and separate their waste, 67% introduced processes to reduce their environmental impact, 53% introduced a new product or service with a specific social or environmental element, and 67% started training their employees on sustainability issues.

   These are great achievements, without any extra large expenditure. Over 90 % of the participating suppliers indicated that the integrated responsible business practices had a positive impact on their relation to their customers and over 80% said that those practices are likely to improve their relationship with other companies in their supply chain.

CSR & Business Areas

Apart from industries, other areas of business, where CSR and consequently environment is becoming a dominant factor include the followings:

- Investors: are starting to show more interest, *e.g.*, $1 out of every $9 under professional management in USA now involves an element of 'socially responsible investment'.
- Banks: big banks, like Goldman Sachs and UBS, have started to integrate environmental, social and governance issues in some of their equity research.
- Finance Industry: mixed signals of giving emphasis on good financial results notably on private –equity, but partly skeptical on CSR issues. But its private-equity itself is to respond to public pressure by agreeing on voluntary codes of transparency
- Employee's demand: Firms are facing strong demand from their own employees on CSR. CSR efforts help to motivate, attract and retain talents in the company.

Summary

i. CSR is a balanced approach for organizations to address economic, social and environmental issues in a way that aims to benefit people, communities and society. CSR has led to induction of environment and community in corporate culture.

ii. CSR includes voluntary commitments by companies. CSR guidelines include OECD -2000 Guidelines on CSR, UN global compact (based on Rio summit), ILO, as well as standards, such as, ISO 14001 (for the environment) and SA 8000 (for human rights) or ISO 2600 (for social responsibility).

iii. Rio Earth Summit dealt with the voluntary protection of environment. Kyoto Protocol defines legally binding targets and timetables for cutting the greenhouse-gas emissions of *industrialized countries* (excepting USA, not ratified KP) that ratified the Kyoto Protocol.

iv. However, corporates' misdeeds, have led to non-governmental groups calling for new laws to protect people and the environment from the impacts of big business, which would make CSR as mandatory.

v. There has been a paradigm shift on CSR activities. CSR is no longer considered as a loosing proposition, but a positive endeavor for prosperity and well-being of both the corporations & the societies in which they belong

REFERENCES

1. CSR in Wikipedia
2. *The Economist*, A special report on CSR, p. 3-22, 19th Jan., 2008

2a. Al Gore; *An Inconvenient Truth*, Rodale, NY, 2006

3. UN-Global Compact.http://www.unglobalcompact. org/ AboutTheGC/ /index. html
4. *The Economist*, January 19th,'Going Global'p17-18.
5. *The Economist*, 9th February, 2008, p13; letter to editor
6. Changing Landscapes: Towards a Sustainable economy in Asia, Conf., co- organized by UN Global Compact, UNEP Finance Initiative (UNEP FI), UN-backed Principles for Responsible Investment (PRI), Seoul, N Korea, June, 2008, http://www.csrwire.com/News/12408.html
7. Tony Juniper, *The Guardian*, Monday, April3, 2006
8. Green CSR in Murata http://www.murata.com/ corporate/csr / management / *compliance /index.html*
9. Shuichi Yamamoto, Murata, *Prevention of Global Warming*, http:// www.murata.com/ corporate/csr/ report/2007/pdf/p31.pdf
10. A Change in Climate - The greening of corporate responsibility, *The Economist*, Jan19th, 2008, p. 13-17.

Bibliography

Berry, R.G. & R.J. Chorley, (2002). *Atmosphere, Weather, & Climate*, Solar Radiation and Global Energy Budget, p20-50, Routledge, NY.

Berry, Roger G. & Richard J. Chorley, (2002). *Atmosphere, Weather & Climate*, Routledge, London and New York, 7th edition, reprinted in 2002, pp. 1 -18.

Brown, Sandra. *Present and potential roles of forests in the global climate change debate*, US Environmental Protection Agency, Western Ecology Division,Oregon, USA,http://www.fao.org/ docrep/ w0312e/w 0312e03 .htm).

Campbell, N.A. (1996). *Biology*, 4th edition, The Benjamin/Cummings Publishing Co. Inc., Menlo Park, CA, USA.

Chapman, J.L. & M.J. Reiss, (2000). *Ecology-Principles and Applications*, 2nd edition, Cambridge University Press.

Chatterjee, Mandira (2010). Urban Heat Island and Climate Change; National Seminar on Dynamics of Cities and City-Regions, Development and Environmental Issues in Contemporary India, 20-22 January, organised by Dept of Geography, Mumbai University, & The Bombay Geographical Association.

Chattopadhyay, R. (1993). High Silt Wear of Hydro-Turbine Runners: *Paper presented at 9th. Int. Conf.on 'Wear of Materials'*; San Francisco, USA, 13 – 16 April, published in the proceedings by Elsevier Publication

— (1993). Rail Wheel Contact Wear, *Intl. Symp. On Tribology*, 18-23 Feb, 1993, Beijing,China, National Science Foundation of China,Chinese Mechanical Engineering Soc., and Tsinghua University.

— (1995). invited speaker, Reconditioning of Thermal & Hydro Power Plants-Indian Experience: Indo-French Seminar on Efficient Management of Power Plants,16-17feb, Taj Hotel, Bombay.

— (2001). *Surface Wear-Analysis, Treatment and Prevention,* ASM International, Materials Park, OH, USA.

— (2004). *Advanced Thermally Assisted Surface Engineering Processes,* Kluwer Academic Publishers, MA, USA, (Now Springer, NY).

— (2009). invited talk on *'Surface Engineering to Control Global Warming', Surface Engineering Seminar,* sponsored by Indian Institute of Metals, Tata Research and Development Centre, Pune, 23 January.

Choate, William T. (2003) Energy and Emission Reduction Opportunities for the Cement Industry, December 29. Energy Efficiency and Renewable Energy, US Department of Energy

Elton, C. S. (2001) *Animal Ecology,* 1927. Republished 2001. University of Chicago Press.

Fan, Zhen, Steve Goidich, Archie Robertson and Robert Roche, *Performance and economics of ultra supercritical pressure CFB boiler power Tplant,* Foster Wheeler North America Corp., Livingston, NJ 07039,http: //www.fwc.com/publications/tech_papers/ files/TP_CFB_07_04.pdf

Hunt, Margaret W (2008). Energy Independence, editorial, Advanced Material & Processes, ASM-International, July.

Juniper, Tony (2006). *The Guardian,* Monday, April 3.

Kemp, David D. (2004). *Exploring Environmental Issues –An integrated approach,* Routledge, London & New York.

Khrushch, Marta & others, (1999). *Carbon Emissions Reduction Potential in the US Chemicals and Pulp and Paper Industries by Applying CHP Technologies,* Proceedings of the 1999 ACEEE Summer Study on Energy Efficiency in Industry, Saratoga Springs, NY, June 15-16,

Kunzig, Robert & Wallace Broecker, (2008). *Fixing Climate –The story of climate science –and how to stop global warming,* Profile Books Limited, London.

Larsen J.M. & others (2002). *Achieving the Potential of Materials Prognosis for Turbine Engines*, DARPA Bidders Conference on Materials Prognosis, 26th September.

Lisa A. (2008). Swenarski de Herrera, *Span Magazine*, March-April, p14-17 geothermal from bauxite.

M. Pidwirny, (2006). The Nitrogen Cycle, *Fundamentals of Physical Geography, 2nd Edition.* http://www.physicalgeography.net/fundamentals/9s.html

Murata, Shuichi Yamamoto, (2007). *Prevention of Global Warming*, http://www.murata.com/ corporate/csr/ report/2007/pdf/p31.pdf.

Prakash, Vandana (2009). *India one of the least Carbon Intensive Countries in the World*; McKinsey Reports, Published on May 24th.

Robinson, Peter J. & Ann Henderson-sellers (1999). *Contemporary Climatology*. Chapter 2, The Earth's Radiation Budget, p17 -38, Longman, 2nd edition.

Robinson, Peter J. & Ann Henderson-sellers (1999). *Contemporay Climatology*, second edition, Longman.

Roger, Revelle (1966). *The Scripps Institution of Oceanography*, "The Role of the Oceans." *Saturday Review*, 7 May, p. 41, http://www.aip.org/history/climate/co2.htm.

Saha, Meghnath (1920). *On Ionization in the Solar Chromosphere*, Philosophical Magazine. http://www.encyclopedia.com/topic/Meghnad_Saha.aspx

Samant, Arvind (2011). Knock on Wood, *Hindustan Times*, Mumbai edition.

Schnare, David (2007). *Environment & Climate News*, The Heartland Institute.

Sen, P.K., (2011). Guest editorial, *IIM Metal News*, vol14, no, 3, 3 June.

Skinner, B. (2000). *The Dynamic Earth*, page 21. John Wiley & Sons, Inc.

Tennsen, Michael (2004). *Global Warming*, Alpha Books, NY, USA.

Weedy, Todd (2009). Alternate Energy Projects Stumble on a Need for Water, *The New York Times*, September 29.

Index

A

Absorption 21, 41-42, 44, 46, 50-54, 56, 60-61, 64, 67, 70, 73-74, 77-79, 82, 88, 103, 106, 112, 118, 121, 145, 147, 150, 175-76, 192, 195-97, 199-200, 264-65, 279, 282, 284
Abyssal zone 102
Afforestation 151, 160-61, 257, 272-74, 277, 288, 291, 314
Agricultural waste 223
management 287
Airlines Green Program 257
Alcoholic fermentation 221
Alpine tundra 116
Alternative energy 307
Amazon rainforest 117
Ammonia production 193, 236
Anaerobic fermentation 221, 290
Anthropogenic climate change 60, 72-73
greenhouse gases 13, 46, 72, 78, 191
Anthrosphere 2, 121, 123, 125-27, 141, 148
Arctic 34, 96, 103, 106, 108-09, 115-16, 118
tundra 115-16
Asahi Glass Co. 338
ASTM D6866 177-78
Atlas of Biomes 119, 285
Aurora 15

B

Bali conference 157, 163-65, 173, 268-69, 273
Bauxite residue 198, 244
Bergeron-Findeisen process 26
Bio-chemical 221
Biodegradable waste 220, 302
Biodiversity 2, 114, 116-19, 153, 168, 273, 275-76, 285, 323-24, 330
Biofuel production 220, 232
Biofuels 153, 203-04, 210, 219-24, 226, 229, 232, 239, 267
Biogeochemical cycling 278
Biological diversity 158, 267
pump 103, 280-81
Biomass burning 86, 287

energy 210, 274
Biomes 2, 31, 33, 73, 95-97, 99, 104, 107, 110, 115, 118-19, 126, 145-48, 153, 155, 157, 263, 276, 284-85, 297, 300
Biorefinery 211, 225, 239, 304
Biosphere 2, 7, 121-29, 139-41, 148, 181, 264
Black liquor 239
Blue jets 13
Boreal forests 113, 271
Boyle's and Charlie's laws 17
Burning fossil fuels 3, 73, 140, 176, 216
Business areas 341
reaction 317

C

California Global Warming Solutions Act 314
Carbon Banking 318
budget 265
conservation 266
credit 3, 152, 176, 178, 183-85, 243, 277, 282, 307-10, 313, 315, 319, 321-22, 330, 334
cycle 2, 42, 53, 78, 107, 127-28, 132, 137-39, 142, 175, 264, 268, 275, 279, 280, 285
dioxide 1-3, 7-8, 21, 42, 53, 56-57, 59-65, 67, 71-74, 77-83, 88-89, 92-93, 98-99, 103-07, 112, 118, 123-27, 137-40, 146-50, 152-55, 159-63, 167-69, 173, 175-76, 179-81, 183, 185, 188, 191-93, 195, 197-200, 204-06, 208-10, 221, 226, 229, 233-34, 236, 238, 240, 242-46, 249-55, 258-59, 268, 272, 277, 279-83, 291-92, 302, 308, 311, 314, 316-18, 323, 340
dioxide concentration 61, 64-65, 88, 139, 175
dioxide emission 53, 57, 107, 149, 153, 159, 167, 173, 175-76, 180, 188, 192, 205-06, 209, 229, 233-34, 238, 240, 242, 249-52, 257, 259, 277, 291, 316
emission 155, 157, 163-65, 171, 176-77, 181, 184, 187, 190, 221, 223, 228, 230, 235, 239, 254-55, 266-68, 272, 292-93, 304, 318, 321-22, 330, 334, 335, 339-41
exchange 265-66, 279-80
flux 264-66, 281, 284, 288
footprint 152, 154, 165, 177, 181-82, 190, 236, 299, 308, 317-18, 326, 339, 341
intensity 168, 186, 188
management 175-76, 189, 266
management mystem 175-76
mapping 275
market 164, 277, 308, 310, 315, 317
neutral 181, 183-84, 189, 204, 210, 224, 238, 264, 274, 292, 294, 314, 319, 321, 325, 339
offset management 177, 182
pool & fluxes 285
pool 264, 266-67, 280-81, 284-85
savings 294

sequestration 191-92, 199, 201, 267, 273, 275, 277, 290-91, 294-95
sink 103, 114, 191, 264, 277-78, 284, 291
storage management 267
substitution 267
thinking 189
trust 318
unit 179, 183, 307-08
ranching 271
Changes in crop rotations 291, 294
Chicago Climate Exchange 277, 313, 316
climate market 277
Circulation cells 23, 27, 39, 146
Clean development 166, 186, 209, 243, 304, 309-10, 312-13, 315, 319, 320-21
Climate change 1-2, 27, 34, 39, 42, 59, 60-61, 66-69, 71-73, 82, 84-85, 89, 91, 94, 96, 117, 127, 131, 133, 140, 157-58, 160, 163, 165-66, 168, 171-73, 180, 223, 228, 238, 242-43, 246, 255, 266, 268, 272-73, 275, 278-79, 285, 292, 317, 319, 323-24, 326, 330-31, 333
model 46, 83
zones 2, 23, 30-33, 41, 95
savanna 113
Cloud forest or fog forest 111
Coastal 62, 105-06, 109, 116, 131, 137, 282-83, 290
and Cold 107
Composition 15, 22, 54, 57, 123, 126-27, 267
Conservation of Biomes 263
Conversion of cropland to pasture 292, 294
Copenhagen conference 173
Coral reefs 100
Coriolis effect 27, 29, 35
Corporate social responsibility 154, 323-24, 326, 332
Costa Rica 111, 225, 232, 267, 285
Country's Emissivity 188
Crop and Garden Production 289
Cryogenic fractionation 195
Cyanobacteria 104, 129
Cyclone 24, 34-35, 40

D

Declaration, Rio 158
Definition 72, 81, 90
Deforestation 3, 55, 73, 88, 110, 116-17, 125, 127, 128, 145-46, 157, 161, 163-64, 166, 173, 175, 213, 224-25, 264, 266, 269, 271-74, 276-77, 284, 287-88, 314, 323
Demand for meat 271
Denitrification 129, 130-31
Density 14, 17-19, 114, 117, 134, 208, 224, 266-67, 276, 283, 290
Derived savanna 114
Deserts 28-29, 39, 54, 96, 107-09, 119, 217
Disasters 1, 8, 24, 34, 39-40, 69, 73, 324

Doppler Radar 20
Dry conifer forests 112
Dynamic entities 2, 96

E

Earth's Atmosphere 1, 7-8, 14-15, 19-23, 26, 28, 43, 46-47, 50, 56, 88, 91, 146, 216, 272
Ecosystem 2, 30, 34, 38, 95-96, 99-100, 104, 117-18, 122, 124, 130-32, 137, 265, 267, 276, 278, 314
Ecotourism 278
Eddies and cyclones 106, 281
Eddy Correlation Flux Measurement 266, 285
El /Nino Southern Oscillation 36
Electron density 14
Emission allowance system 184
  data 176-77, 233
  factor 186-88
  intensity 186, 188
  limits 149, 175-76, 184, 252, 260
  management 154, 184, 233, 235-36, 250, 287, 309, 323
  trading 172, 179, 182-83, 185, 246, 307-13, 319, 321
  trading scheme 185, 311, 319
Energy consumption 3, 151, 176-77, 180, 182-83, 189, 204, 230, 234-35, 242-43, 245-47, 321, 338
  processing efficient 235
Enteric fermentation 154, 162, 288-89, 294
Environment 26, 66, 69, 80, 94-95, 116, 119, 124, 137, 147, 158, 160, 165, 168, 174, 181, 183, 189, 201, 231-32, 261, 264, 272, 275, 283, 285, 293-94, 297, 300-01, 305, 314-15, 319, 322-23, 325-27, 329-30, 333-36, 341-42
Epipelagic 101
Eroded Turbine 213
Estuaries 99-100, 104, 278
Euphotic 105, 282
European emissions-trading proposal 255
Evergreen rainforest 111
Exosphere 9, 14-15, 20-21
Exploitation/logging, timber 271

F

Federal Aviation Administration 254
Ferrell cell 33, 146
Fertilizer 80, 86, 90, 99, 132, 176, 223, 226, 237, 283, 287-90, 293, 302, 304
Field burning 154, 162, 288
Flexible mechanisms 185, 312-316
Forest carbon flux 288
  preservation 268, 269, 277, 307
  products 267, 274
  soils 266
  vegetation 266
Fossil fuel based power generation 204
  fuels 3, 59-60, 65, 73-74, 88, 90, 123, 127, 137, 140, 153, 158,

163, 168, 176-77, 180, 192, 203, 205, 210, 216-17, 220, 222, 230, 240, 245, 249, 263, 264, 272, 284, 287, 292, 333

Freshwater 97-99, 104, 118, 132, 278

G

Geological storage 192-93, 197-98

Geosphere 2, 7, 121-24, 127-28, 141, 148

Geothermal 204, 210-11, 214, 226-27, 229, 232, 244

Global carbon cycle 107, 132, 138, 264, 268, 275

efforts 157

issue 326

sectorial approach 242

warming 1-4, 8, 22, 27, 33-39, 41, 48, 50, 53, 57, 59-61, 65-68, 70, 72-74, 77-78, 80-81, 83-84, 86, 89-91, 94, 96, 99, 104, 116, 118-19, 121, 127, 137, 140, 145-48, 151-55, 157, 166, 172, 175-76, 179, 181, 191, 204-05, 216, 224, 226, 228, 230, 235-36, 244, 255, 268, 274, 279-80, 292, 294, 299, 305-06, 308, 311, 316, 318, 323, 325-26, 336, 343

warming potential (GWP) 81, 83, 92, 166, 179, 308, 311, 316

Green buildings 320

cities 307, 320

climate fund 172

constructions 307

theme 334

Greenhouse gases 1-3, 8, 10, 13, 16, 24, 42, 46, 54, 59, 61, 67-68, 70-72, 77-83, 85, 89, 145, 147-52, 158-60, 162, 173, 176, 178, 181, 184, 187, 192, 200, 215, 224, 226, 236, 240, 253, 277, 317, 320, 336

Guidelines 3, 148, 152, 159, 161, 186-89, 289, 327, 329-30, 338, 342

H

Hadalpelagic Zone 102

Hadley cell 28-29, 146

Hadley's circulation cell 24, 33

Heterotrophs 98

High Erosion Silt 213

Himalayan region 34, 213

Human activities 2, 71, 78, 80, 89, 114, 121, 140-41, 148, 150, 181, 282, 284

Humidity 1, 10, 16, 23-27, 31, 33-35, 39, 56, 85, 108, 136, 148, 279

Hurricanes 24, 34-35, 40

Hydro Power 214, 231, 339

Hydrographic fronts 106, 281

Hydrosphere 2, 7, 121-27, 141, 148

I

India 3, 14, 18, 30, 32, 40, 75, 111, 113, 163, 165-71, 173-74, 186, 189, 204-05, 209, 211-14, 220, 223, 229, 231-32, 236, 239, 241-42, 246-47, 250, 259, 261, 270-72, 275, 293, 295, 304, 311, 321-22

Indian railways 258

scenerio 275
Innovative processes 145, 153, 155
International transaction log 309-11
Intertidal zone 100
Investors 164, 189, 315, 324, 341
Irreversible environmental changes 273

K

Karman line 9, 21
Koppen-Geiger classification 31
Kyoto Protocol 2, 81-82, 85, 93, 149-52, 154, 157, 159-60, 162-64, 169-73, 176-77, 179, 183-85, 187, 189-90, 209, 234, 268-69, 288, 307-13, 315-17, 319-21, 323-24, 330-31, 342

L

La Nino 38
Land degradation 287
use change 55, 273, 287-88, 292, 294
Less-intensive tillage 291, 294
Life Cycle Assessment 182
Limnetic zone 98
Lithosphere 121-22
Littoral zone 98

M

Magnetosphere 14
Mandatory market 176, 308, 312-13
Manure management 162, 288-89, 314
Marine 69, 73, 96, 104, 106-07, 118, 122, 132, 140, 160, 224, 259, 278-79, 282-84, 331
Market price for carbon 315
Market size 316
trend 316
Measurement 20, 47-48, 61-62, 64-65, 74, 83, 164, 170, 184, 190, 259, 265-66, 277, 285, 309, 320
Mediterranean forests 112
Meghnath saha 14, 22
Mesopelagic zone 102
Mesosphere 9, 13, 19, 21
Methane 1, 8, 16, 71-72, 77, 79-81, 84-86, 89-90, 92-93, 124, 127, 140, 147, 161-62, 175, 178, 180, 182, 194, 223, 268, 287-90, 294, 300, 304, 311, 314, 331
Mid-latitude cyclones 30
Midnight zone 102
Mineralization 129
Minerals 103, 105, 110, 118, 122, 138, 141, 206, 233-35, 246, 281, 298
Moist/dry deciduous forest 111
Moisture 10, 21, 26, 38-39, 50, 79, 85, 91, 95, 108, 115, 132-34, 146
Murata Group 335-36
Mustard stick 223

N

Natural emission 264
spheres 2, 121, 123, 141, 148

Nimbus 25

Nitric Acid production 236

Nitrification 129-30, 290, 302

Nitrogen 7-8, 14-16, 20-21, 50, 78, 107, 121-22, 124, 127-32, 136, 141, 146, 148, 178, 187-88, 193-94, 289, 302

Nitrous oxide 1, 8, 26, 71, 79, 80, 84, 90, 93, 124, 130-31, 180, 187, 233, 236-37, 248, 255, 287-88, 294

No-till soil management 292

Nuclear power 161, 227-28, 230, 232

O

Ocean area 107
  circulation 27, 69, 278
  Storage 198

Oceans 2-3, 62, 66, 69-70, 73, 80, 97, 99-100, 103-04, 106-07, 122, 125-27, 132, 134-36, 140, 153, 155, 175, 225, 263-64, 278-79, 282-84

Offset projects 257, 273, 320

Oil spills 283

Orbital forcing 54-57

Orbital variation 67, 69

Outer space 9, 14-15, 17, 21, 77

Oxygen 7-8, 12, 14-17, 19, 21, 78, 88, 91, 98-99, 102, 104-05, 107, 111-12, 118, 122, 124, 126, 130-31, 140, 146, 178, 194, 200, 207, 224, 240, 268, 278, 281, 283, 290, 293, 304

Ozone layer 7, 12-13, 19, 21, 33, 44, 50, 56, 92-93, 126-27, 147, 157

P

Papua New Guinea 117, 276

Permafrost 115-16

Photosynthesis 16, 42, 53, 56, 88, 98, 103-06, 112, 118, 122, 124, 126, 137, 139-40, 147-48, 154, 191, 199, 220, 268, 278, 280-82, 292

Photovoltaic effect 218

Physical pump 280-81

Phytoplankton 98, 104-07, 110, 140, 148, 154, 191, 279-84

Planck's Law 44, 46

Planktons 88, 103-06, 118, 281, 282

Plasma 14

Plastic garbage 283

Polar cell 2, 27, 29, 30, 146
  easterlies 30
  vortex 30

Ponds and lakes 97-98

Precipitation 1, 10-11, 20, 23, 24-26, 31-33, 39, 46, 69, 80, 109-10, 112-15, 125, 131, 133-36, 138, 146, 279-80

Pressure 1, 7, 9-11, 17, 20-21, 23-24, 26-29, 34-36, 38-39, 46, 79, 102, 129, 146, 193, 196, 208, 231, 242, 261, 280-81, 303, 334, 342

Prevention, Pollution 94

Primary boundary layer 9
  consumers 105

energy savings potential 242
producer 105, 124
Profundal zone 98
Protected area system 117, 267, 276
Pyroprocessing 245

R

Radiative efficiency 81-82
forcing 54-55, 57, 67-68, 70-72, 81, 82, 84, 94, 180
Radio waves 14
Radiosondes 19
Rainforest 32, 111-12, 117, 127, 278
Relative humidity 16, 24-25, 27, 85
Renewable energy certificates 314-15
resources 151, 210, 215, 228, 234-35, 334
Residue management 291, 294
Rice cultivation 154, 162, 288, 290
Rio earth summit 328, 330, 342
Road building 271-72
Ruminant animals 288-89

S

Seasonal rainforest 111
Seasons 12, 24, 30, 37, 39, 69, 108, 110, 112-13
Secondary consumer 105
Seeding 107, 118, 154, 282, 284, 293, 319
Semi evergreen forest 111
Semiarid 31, 107-09
Semiconductor 91, 218-19
Sewage disposal 283
Silviculture 267
Skin tusk 271
Slash and burn farming 271
Soil management 288, 292
tillage 287
Solar cycle 48-49
irradiance 42, 47-50, 54, 56-57, 216
luminosity 42, 48-49, 54, 56-57, 60
radiation 8, 17, 19, 21, 26, 41-48, 51, 53-57, 70, 79, 91, 108, 111, 118-19, 151, 191
radiation management 191, 200
thermal 211, 214, 217-18
Storage of carbon 132, 155, 192, 267
Storm Fronts 34
Stratosphere 7, 9, 11-13, 17-19, 21, 50, 86, 91-93, 126, 147, 191
Streams and rivers 98, 138
Subtropical high pressure zone 29
Sulphur dioxide 8
Sun's Surface 42, 49

T

Temperate forest 112-13
grassland 113-15
Temperature 1-2, 7-11, 13, 15-28, 31, 33-34, 37-39, 41-42, 44-47, 49, 54, 57, 59, 60, 67, 71-74, 77-79, 80, 82-84, 88, 91, 93, 98, 101-02, 106, 108-10, 113-14, 116,

118, 124-26, 129, 135-37, 145-48, 194, 199, 207-08, 216, 224, 227, 229, 237, 256, 266, 279, 280, 281, 283-84, 303

Terrestrial 45, 54, 96, 113, 117, 122, 263, 264-66

Thermal lows 28

Thermal storage 218

Thermocline 36, 98, 101

Thermosphere or ionosphere 13

Tornadoes 24, 26, 34-35, 39-40

Toxic chemicals 283

Transmission grid 215, 217

Tropical 2, 24, 29, 32-36, 39, 68, 73, 89, 96, 98, 106, 110-13, 116-17, 125, 153, 164, 264, 266, 269-71, 274, 276, 282

Tropical forest 110, 113, 270

Troposphere 7-12, 15-19, 21, 23, 26, 28-30, 50, 70, 86, 89, 126, 146

turbulent 9-10, 35, 207

Turbulent mixing 9

Twilight zone 102

U

UN global compact 327-28, 330, 342-43

Upwelling 36, 84, 106, 118, 281

Urban Heat Island 72, 75

UV rays 44

V

Vertical mixing 10-11, 19

Vietnam 117, 276

Voluntary market 176, 307-09, 313-15, 321-22

W

Walker circulation 36

Waste energy recovery 242

management 3, 155, 161, 228, 297, 300-05, 331, 334, 341

reduction 337

Water vapor 7-8, 10-11, 15-17, 21, 23-26, 39, 42, 68, 77, 79-80, 83, 85, 123, 125, 134, 136, 138, 148, 266, 280

Weather 1, 7-10, 16, 18, 20-24, 27, 34, 35, 37, 39-40, 50, 57, 60, 66, 73, 94, 107, 114, 137, 146

radar 20

Western Climate Initiative 314

Wetlands 86, 97, 99, 291

White Coal 223, 232

Wind power 153, 210-11, 215-16, 229, 230

Window of opportunity 272

World Energy Conference 157, 163

WWF efforts 277